Basic
Color
Television
Course

By Stan Prentiss

TAB BOOKS

Blue Ridge Summit, Pa. 17214

FIRST EDITION

FIRST PRINTING—APRIL 1972
SECOND PRINTING—JULY 1974
THIRD PRINTING—JUNE 1977

Copyright © 1972 by TAB BOOKS

Printed in the United States
of America

Hardbound Edition: International Standard Book No. 0-8306-2601-8

Paperboudn Edition: International Standard Book No. 0-8306-1601-2

Library of Congress Card Number: 73-189960

Contents

& IC Color Subsystem—Motorola's Tri-Phase Chroma
Demodulator—Zenith's Three-Chip Chroma Processor—
RCA's Two-Chip Chroma Processor—A One-Chip Complete Monolithic Chroma Processor—Questions

Preface

This book is a product of 25 years in the TV business, talks with hundreds of engineers and technical people from the entire television industry, beginning with representatives of most of the major receiver manufacturers, the FCC, the CATV people, TV broadcasters, semiconductor manufacturers, wholesalers and retailers, and service personnel from every strata. Deliberately, the book begins with simple AM and FM modulation, on the theory that a person who doesn't know these basics couldn't possibly understand how transmitters and receivers work. Then, the content discusses the composite color television signal as it is originated at the camera and broadcast from the TV station. From there you can follow it to the receiver antenna, down the transmission line and into the receiver input. In preparing the material, every attempt was made to include the most useful information—not only what is current now, but what was true five or six years ago, and what we know will happen in the forseeable future.

All types of EIA monochrome and color test patterns are discussed, along with the new vertical interval test signal (VIT) and the proposed vertical interval color reference signal that could eventually automatically control color tint and amplitude (VIR). There is a presentation on antennas, on transmission lines, on the newest 110-degree color picture tubes, an entire chapter on cable television and foreign television systems (PAL and SECAM), and a very careful analysis of almost every possible circuit used in major tube, discrete semiconductor, and integrated circuit (IC) television receivers produced to date. At the end of each chapter, there are a number of review questions to help you test your knowledge of the facts presented. Answers to the questions appear at the back of the book.

Stan Prentiss

Acknowledgments

In writing this book, we've had much experience, a great deal of unselfish, useful help, and a wealth of information from which to extract, expand, and offer you in what should be the most comprehensive analysis of color television—and perhaps monochrome, too—available today. To all those grand people—and we unfortunately haven't the physical space for titles—who have helped so much, this book, then, is gratefully dedicated:

Tektronix—Dan Guy, Marshall Pryor, Jim Walcutt
Zenith—Al Cotsworth, Jane Temple, Ed Kobe, Frank Hadrick, Ed Polcen
RCA—Tom Bradshaw, Don Peterson, Dick Cahill, Dick Santilli, George Corne, Ed Milbourn, Carl Moeller, Charles Meyer, Dave Carlson
Motorola— Ralph Jones, Bill Slavik, El Mueller, Chuck Preston, Stan King, Sid Schwartz,
Fairchild—Norman Doyle, Don Smith, Bill Edlund
EIA—Gene Koschella
JFD—Jim Sarayiotes
Delco—Jim Hornberger
G.E.—Bob Hannum, Frank Boston
EXACT—Jerry Foster, Wayne Hunter
NBC—John Platt, Jaro Lichtenberg
Rauland—Jerry Ridgeway
Sylvania—Joe Thomas, Gene Nanni, Karol Siwko
Hewlett-Packard—Jack Molcan
Jerrold—Caywood Cooley, Jr.
Sencore—Herb Bowden, Jim Smith
B & K—Carl Korn, Harold Schulman
Systron Donner—Ed Phillips
Mercury—Harry Rich, Stanley Abrams
Bell & Howell—Mike Laurance, Bob Lundquist

Chapter 1

The Complete Color System

Electrical energy entering free space does so in the form of electromagnetic waves. It travels at the velocity of light, a speed of 3×10^8 meters per second. Collectively, such energy is called a radio wave. A radio wave consists of both magnetic and electric fields that are at right angles to one another and also at right angles to the direction they travel. If these waves are generated by an alternating current, they will vary in intensity and swing both positive and negative as they alternate throughout each cycle. The wavelength of each cycle (lambda), then, is equal to the free-space velocity (rate of travel) divided by the number of cycles transmitted each second:

$$\text{Wavelength } (\lambda) = \frac{300,000 \ (km/sec)}{freq \ (khz)}$$

All operating electrical circuits radiate to some degree. A power line, for example, with relatively short footage spacing between conductors will radiate very little if it is in good working order, since a wavelength at 60 Hz is over 3,000 miles—a huge difference by comparison. But a large transmitter with 50,000 watts output will propagate enormous energy since it uses a large radiator with a length inversely proportional to frequency. Therefore, high-frequency waves may be broadcast over considerable distances by medium and large-size radiators. The limiting factor in very high frequency transmissions, such as TV, is called the "line of sight." Such transmissions are not affected by the ionosphere, and so will go directly off into outer space at the horizon—a characteristic that has been made good use of in many forms of aircraft and satellite communications, as well as microwave.

For a transmitted wave to contain intelligence, some part of it must be varied (modulated) in time and degree (frequency deviation or amplitude) by the information you wish it to carry. In radio broadcasts, this variation is achieved

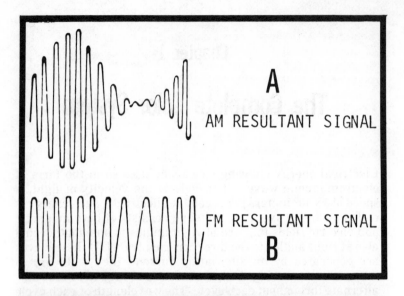

A

AM RESULTANT SIGNAL

B

FM RESULTANT SIGNAL

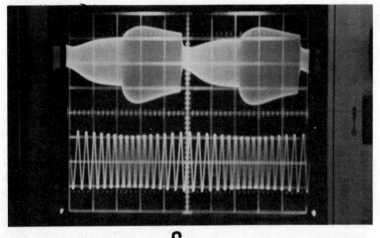

C

Fig. 1-1. Drawing of an amplitude (AM) modulated waveform, where the envelope amplitude is varied by "contraction" and "expansion" (A). Frequency modulated waveform (FM) showing carrier frequency deviation but not amplitude variation (B). Scope waveforms showing 400-Hz AM modulation (top) of a nonlinear sine-wave generator operating at 12.5 MHz. FM modulation (bottom) is represented by a 1.6-MHz carrier modulated by a 100-kHz signal.

either by amplitude (AM) or frequency modulation (FM). And in television transmissions, the sound and picture intelligence is dispatched by both FM (sound) and AM (picture). Also, because of the various pulses for vertical and horizontal picture synchronization (sync) and another pulse for color synchronization we almost have **phase** and **pulse** modulation in addition. So in color television, many of today's electronic marvels are collected into a single unit.

AMPLITUDE (AM) AND FREQUENCY (FM) MODULATION

The beginning of almost all signals at RF frequencies is a sine-wave (an alternating wave) generator that oscillates at the **mean** (middle) frequency of any designated RF band. The generated wave is called a **carrier**, and what is done with this carrier determines the information that it will transport from some point of origination to another point of reception.

In **amplitude modulation**, for instance (Fig. 1-1A), the entire envelope of the carrier amplitude is "expanded" or compressed by the frequency and variation (with time) of another wave representing sound, picture, etc. If such a modulating signal had a 1-kHz (1,000) frequency, the higher frequency carrier would be subjected to 1,000 amplitude variations each second.

In **frequency modulation** (Fig. 1-1B), the carrier amplitude remains constant while the carrier frequency itself is varied (contracted and expanded) according to both the intensity (magnitude) of the modulating signal and its rate (time). Therefore, if the same 1-kHz signal modulated a 1,000-kHz sine wave carrier, the deviation could range from 1,000.1 kHz to 999.9 kHz on small output signals to 1,010 and 990 kHz on large signals. This analysis immediately suggests there could be a number of sine waves modulating a carrier with pairs of frequencies surrounding the carrier frequency. And this is true, since, for every frequency apparent in any modulating wave, there are two immediate side frequencies generated, one above and the other below the main carrier. These are called the **upper** and **lower** sidebands and are especially important in the study of color television.

THE NTSC COLOR SYSTEM

As with monochrome, the allowable frequencies for the VHF channels begin with 54 MHz for Channel 2 through 88 MHz for Channel 6. There's then a skip from 88 to 108 MHz for

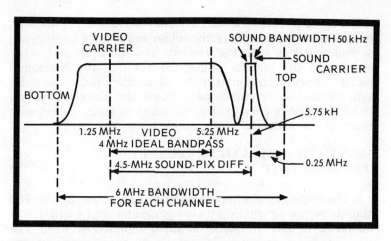

Fig. 1-2. Drawing of the 6-MHz bandpass response for all VHF-UHF TV channels showing the 4.5-MHz difference between sound and picture carriers and an ideal video bandpass of 4 MHz, something that's not realized with color.

the FM-stereo band, and from 108 to 174 MHz for other types of services. Channel 7 begins at 174 MHz and the TV spectrum continues through Channel 13 at 216 MHz. UHF stations start with Channel 14 at 470 MHz and end on Channel 83 at 890 MHz, a total of 70 UHF spaces in the spectrum.

Each VHF-UHF channel has a bandpass (bandwidth) of 6 MHz (Fig. 1-2), with the video carrier 1.2 MHz higher than the lowest frequency specified for an individual channel, and the 50-kHz wide sound carrier positioned 0.25 MHz from the top of the channel, thus separating audio and video by 4.5 MHz. For that reason, you can determine tuner response easily with a sweep generator, a video carrier marker, and a 4.5-MHz marker (for sound). As you'll see later, you can determine both tuner bandwidth and waveform rolloff or symmetry by the position of the markers. The lower sideband width is limited to 1.25 MHz to prevent interference with an adjacent channel, but the upper sideband extends a full 4.25 MHz above the picture carrier.

Constriction of the lower sideband to 1.25 MHz is called **vestigial** sideband transmission and helps keep the overall channel passband within the allotted 6 MHz. The transmission is made possible by a carefully designed, fairly sharp rolloff filter that attenuates but is accurate within the limits of the carrier frequency because its edge coincides with the carrier frequency. Fig. 1-3 shows this lower sideband attenuation and also puts the picture carrier in immediate perspective.

In Fig. 1-3 the I and Q color sideband signals have been added, plus the color subcarrier frequency of 3.579,545 MHz—the carrier for all color information. The I and Q sidebands have the phase relationship to burst signals shown in Fig. 1-4. The burst is nothing more than the color sync signal, and is simply a sample of the original 3.58-MHz subcarrier inverted 180 degrees. From Fig. 1-3, you can see that color information is contained in the I and Q sidebands on either side of the single subcarrier.

I and Q information modulates the 3.58-MHz subcarrier, with I and Q signals each allotted a double sideband position on either side of the carrier from 0 to 0.5 MHz about the carrier. Single sideband transmission also exists from 0.5 to 1.5 MHz. This prevents crosstalk interference between the two color signals when received. We won't talk extensively about the I and Q color intelligence at the moment, but Fig. 1-5 should help explain what is meant now. However, you should understand that the I sidebands produce colors from bluish-green to orange, and the Q sidebands supply information ranging from yellowish-green to purple (magenta). **They both, obviously—and this is highly important—have green components, as does the luminance Y signals**, and this characteristic can be used to advantage both in transmission and reception as you will soon see.

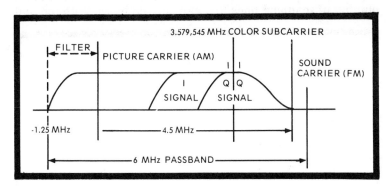

Fig. 1-3. The same monochrome characteristics in Fig. 1-2 are indicated here, but with color information now added. The drawing also shows the effect of a filter for negative vestigial sideband compression. Notice I and Q signals on either side of color subcarrier, plus the color sync sub-carrier.

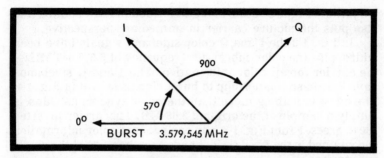

Fig. 1-4. Vector diagram showing the relationship between burst (3.579,545-MHz subcarrier) and the I and Q
sidebands.

THE VERTICAL, HORIZONTAL SYNC & BLANKING SIGNALS

Fig. 1-6 is probably the most copied pattern in all
television instruction books and, perhaps, the least understood. It contains the standard color television sync signals
for video transmission timing in the United States, and **every
one of the individual pulses shown in this figure are broadcast
only during the vertical and horizontal blanking intervals.** The
horizontal blanking interval amounts to 11.1 microseconds,
and the vertical blanking interval measures 1.4 milliseconds,
or 1400 microseconds. Since the vertical blanking pulses occupy about 8 percent of the 525 scanning lines, the maximum
number of scanning lines that can be used is 485. With top and

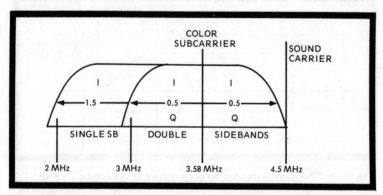

Fig. 1-5. Color I and Q double sidebands are shown on
either side of color subcarrier at 0.5-MHz frequencies,
with I extending to 1.5 MHz in the single sideband (to the
left). The sound carrier is at upper right of diagram.

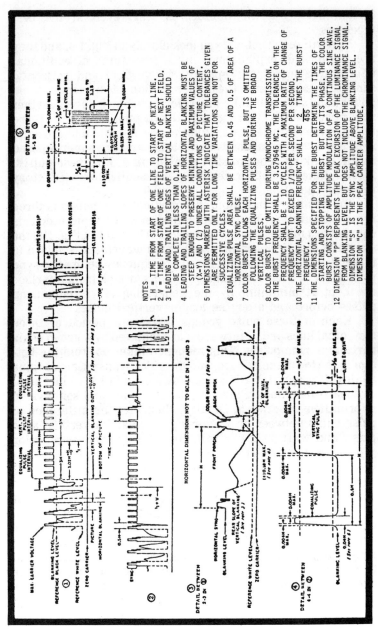

Fig. 1-6. Drawing showing the standard U.S. color TV sync signals transmitted and received only during vertical and horizontal blanking periods when there is no picture.

bottom overscan, the number is less, even in the best receivers.

You will see 12 notes if you look closely in Fig. 1-6, and they are mostly self explanatory, but a few words of additional information seems appropriate. All vertical and horizontal sync tips extend into the upper region of transmitted information, and well into the reference black level, further insurance that they will not normally be seen at any time there is a picture on the screen. Pulses used for line or horizontal sync are 4.76 microseconds in duration and are superimposed on the blanking pedestal shown in Detail 5. For these pulses, fast risetimes and accurate timing is essential for satisfactory receiver operation.

On the other hand, a burst of vertical sync pulses is three times the duration of a horizontal line, or 3 x 63.5 microseconds equals 190.5 microseconds. At the left of Detail 1 in Fig. 1-6, the standard (but here compressed) video waveform is shown in three cycles at the line (horizontal) frequency before the sync period starts. Field frequency problems arising from interlace are kept at minimum by the generation of six equalizing pulses before and after the longer vertical pulses.

Each of these pulse groups last for a period of three horizontal lines and, afterwards, are followed by a group of six horizontal sync pulses. This sequence takes us from the bottom to the top of the picture, and the video transmission begins again. More sync intervals are shown in Detail 2, while Detail 3 repeats the blanking level, horizontal sync, and some video information in between. Detail 4 illustrates the difference in pulse width between an equalizing pulse and a vertical sync pulse, and detail 5 shows a minimum of 8 cycles of burst color sync on the back porch of the horizontal sync pulse that is the second portion of the horizontal 11.1 microsecond blanking interval.

The total series of pulses shown in Detail 1 would take just over a millisecond to complete. Remember, the entire vertical blanking interval is only 1.4 milliseconds. When looking at the composite video signals on a scope set at either the line or field frequency, you will find the two signals rather similar. But at 20 microseconds per division, you can see only two or three cycles of horizontal line information, and at 5 milliseconds per division only 3 cycles of vertical picture information. Scanning or sweep rate is the big difference.

ANALYZING THE COMPOSITE VIDEO WAVEFORM

Since it's much easier to look at a composite video waveform from the receiver (and it should approximate that sent out from the transmitter) let's look at the actual waveforms as you will always see them with a good oscilloscope, then at a line drawing of the composite signal to identify each portion of the waveform. The waveforms in Fig. 1-7 were taken at vertical and horizontal (top and bottom) sweep rates. Notice the upper trace can be counted for 16.66 milliseconds at a 2-milliseconds per centimeter sweep, while the bottom trace, at a 10 microsecond per centimeter sweep, can be counted almost precisely for 63.5 microseconds between sync pulses. In the bottom trace, notice that the 3.58-MHz burst signal sits squarely on the back porch of the horizontal sync signal and pedestal of the blanking pulse where it should be. In signal analysis, you use the peak-to-peak amplitude, scope time base (rate) and DC level to find faults that always seem to occur. The design engineer, too, always uses an oscilloscope to prove his better circuits; many complex equations are often "derived" **afterwards.**

Fig. 1-8 is almost self explanatory, except perhaps for times given for one full cycle at the horizontal rate and another full cycle at the vertical rate. The two blanking intervals are also given for the different waveforms, just as though the trace had a dual time base. Actually, the two traces

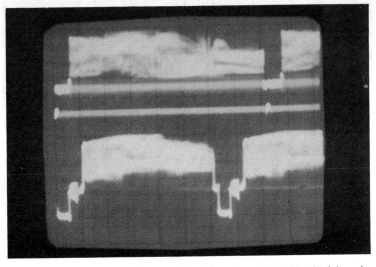

Fig. 1-7. Composite video signals at a 59.94-Hz field rate (top) and a 15,734-Hz line rate (bottom). Due to the necessary double exposure (you can't look at both simultaneously with a single time base scope) they are a little smeary.

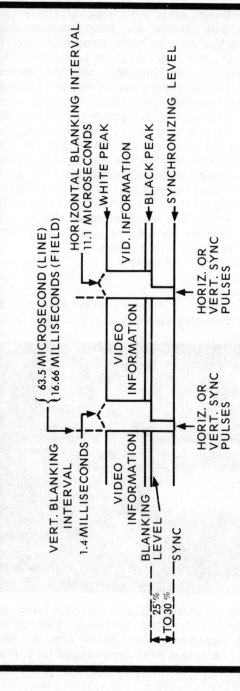

Fig. 1-8. A standard video waveform is very similar in appearance at both line (horizontal) and field (vertical) scope sweep frequencies. The difference is that at the line rate you're only viewing one or perhaps two lines of information, while at the field rate you see the entire raster scanned once for a field and twice (two complete cycles) for a frame (complete picture).

appear very much the same, except that instead of looking at a single line of information at the horizontal rate, you're looking at a single field (one 60th sec) for each 16.6 millisecond interval, followed by the second 262.5 lines if you are showing a second vertical field after the 1.4 millisecond blanking interval. The same is true for the horizontal composite waveform; each full cycle is a new line, and you can have a number of lines in succession, depending on the setting of your oscilloscope's time base.

One additional feature well worth mentioning is the relative black-white-sync levels in the composite waveform. Black is 75 percent of the waveform amplitude; white, 12.5 percent; sync and equalizing pulses range between 75 and 100 percent of the entire envelope amplitude. A knowledge of these proportions will be thoroughly worthwhile later when troubleshooting, modifying, or designing video, AGC, and sync stages of any television receiver.

Color Signal

Basically, we've been talking about monochrome vertical and horizontal sync rates—60 Hz (or 16.6 msec) and 15,750 Hz (or 63.5 microseconds). In color transmission, these frequencies are slightly different—59.94 Hz for vertical and 15,734.264 Hz horizontal—but both are well within the design tolerance of monochrome receivers. The reason for this minute change is to make the color information fit the transmitted waveform envelope and interleave color sidebands with luminance information in the Y channel.

The luminance signal is transmitted as vestigial sideband information and it modulates the video carrier in both color and monochrome transmissions. The I and Q intelligence, however, separately amplitude modulate a pair of 90-degree out-of-phase 3.58-MHz carriers that are suppressed and, consequently, must then amplitude modulate the video carrier. Since the 3.58-MHz carrier is a modulation component of the same Y video carrier, the two color carriers are called subcarriers, and are so referenced throughout the remainder of the book. These subcarriers, of course, have identical 3,579,545-Hz frequencies and similarly position the color information on either side of the fundamentals as illustrated in Fig. 1-5.

To blend the color (chroma) information into the video envelope where monochrome signals don't exist, the altered horizontal frequency is now used directly. Again, in Fig. 1-9 we show the color subcarrier in the central position, as chroma is

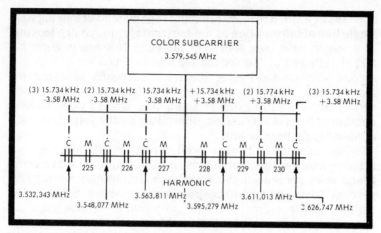

Fig. 1-9. Whole harmonics of the line scanning frequency lie between pods of color information. Both are separated by successive multiples of 15,734-Hz clusters on either side of the color subcarrier. The process is called interleaving, and effectively separates chroma and luminance so both can modulate the same video envelope without undue interference.

interleaved between the various harmonics of the monochrome single sideband clusters. Many books state that interleaving takes place at odd multiples of half the line scanning frequency and this is true, but it's a little hard to visualize. Let's just say that if you add and subtract 15.734, 264 kHz to and from the 3.579,545 MHz subcarrier, the chroma information will be effectively inserted at multiple points that will **not** interfere with the luminance (Y) information. This process is called interleaving. Also, on alternate scan lines, the burst information is transmitted 180 degrees out-of-phase and tends to null (cancel), leaving less interference in the picture.

What's actually happening is due to the fact that the broadcast signal contains many sine waves consisting of harmonics of the line and dual field frequencies with clusters of information gathered about each harmonic at whole—NOT half—multiples of the line and frame (two fields to a frame) frequencies. So if a new carrier with its chroma sidebands is set high in the AM luminance envelope between harmonics of the horizontal scan frequency, there will be little or no interference with brightness (Y) information and the color can be successfully inserted in this envelope at the sum and difference frequencies shown. In the diagram, the two double vertical lines marked M are the monochrome clusters, while each group of three lines with the letters C above are the color pods.

Y, I, & Q SIGNALS

Basically, the luminance signal has a bandwidth of from 0 to 4.2 MHz. The color passbands, of course, are 1.5 MHz for I and 0.5 MHz for Q. Y is produced by the combined outputs of the three color camera tubes and is composed of the following red, blue, and green signal proportions (go two points right for decimal percentages): Y is 0.30 red + 0.59 green + 0.11 blue. Here, the brightness of the entire picture should, at all times, be directly proportional to its color parts, something known as gamma—a correctable condition in the color processing camera amplifier done with shading circuits at a solidly clamped black level. Gamma is defined as the ratio of light to signal and is the numerical exponent (such as Y equals X^2) of any curve describing the light input in lumens to the camera tube output in volts. There is a gamma distortion circuit in the camera processor that actually distorts this relatively linear output signal so that the nonlinear cathode ray tube in any receiver can produce a corrected light-signal ratio of both luminance and chroma. Notice in Fig. 1-10 that the luminance signal is taken off the underside of the RGB matrix whether or not there is an incoming color signal.

The 0 to 1.5-MHz I information is made up of the following proportions of the RGB signal: I is 0.60 red - 0.28 green - 0.32 blue. The Q (0 to 0.5 MHz) intelligence is generated with still other sums and subtractions of the RGB signal in these proportions: Q is 0.21 red - 0.52 green + 0.31 blue. Observe that in the RGB matrix we are combining red, blue and green signals and producing an output of I and Q signals that go to the balanced modulators. When this process is reversed in the receiver, the red, blue, and green outputs are in the following form if the primary colors were derived entirely from I, Q, and Y which, of course, they are NOT, but only because the I and Q information is shifted 33 degrees to R-Y, B-Y, G-Y and then recovered as RGB with the luminance Y signal added. An I-Q recovery receiver would be very expensive, and the RGB-(Y) system gives almost as good results.

The RGB signals are composed of the following proportions: R is 0.94I + 0.62Q + Y, G is 0.27I + 0.65Q + Y, and B is 1.11I + 1.7Q + Y. In monochrome pictures, a small difference in the luminance signal would not be especially noticed. But in color, even a very small error will result in considerable change that would be obvious to any viewer.

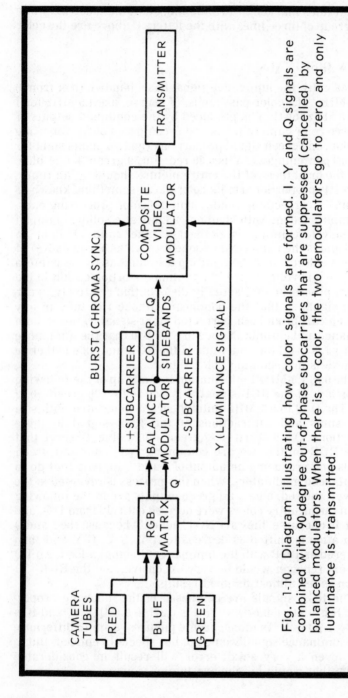

Fig. 1-10. Diagram illustrating how color signals are formed. I, Y and Q signals are combined with 90-degree out-of-phase subcarriers that are suppressed (cancelled) by balanced modulators. When there is no color, the two demodulators go to zero and only luminance is transmitted.

After the matrix changes RGB to I and Q, the two signals are injected into a pair of balanced modulators with 3.58-MHz carriers that are 90 degrees out of phase with one another. These balanced modulators then cancel (or suppress) the out-of-phase carriers and the two-variable I and Q amplitude modulated sidebands are put into the composite video envelope along with burst (for sync) and luminance information for black-and-white detail. The transmitter takes over and puts luminance, monochrome, and color sync signals on the air.

CAMERA TUBES

Cameras and camera tubes are increasing constantly in efficiency and image resolution as the electronic industry develops. The image Orthicon—the standard of color television for so many years—is beginning to be replaced, at least to some extent by the Visticon and the Plumbicon, made by RCA and North American Philips, respectively. The image orthicon has a target material of glass and lacks low light level sensitivity. As its globules lose efficiency, changes take place in the glass target and cause a "sticky" effect that can be seen as image retention when one scene is viewed following another. The new Visticons and Plumbicons have diodes that are considerably more sensitive, making the targets that respond to light focused by the camera lens an orderly sea of small, light-sensitive conductors with a highly uniform output. It's interesting to realize that if one or more of these diodes shorts (and they sometimes do), you can see a pinpoint of light on the studio monitor that immediately identifies this condition. Sometimes the diodes "heal" themselves and resume their normal electronic conduction; but if they don't, then the tubes must be changed. Image orthicons have lasted as long as 40,000 hours, the Plumbicons and Visticons may or may not have as many hours of life; it's too early yet to tell. Visticons and Plumbicons are really lead-oxide vidicons with diode targets—another step along with electronics ladder toward better pictures for television.

THE COMPLETE TRANSMITTER

Fig. 1-10 is a diagram of the complete color part of a transmission system showing how chroma, color sync, and luminance signals are combined before delivery to the broadcast antennas. However, additional signals must be added to the composite chroma and luminance information to produce a stable and complete transmission, along with the necessary sound. Fig. 1-11 includes the remaining elements of

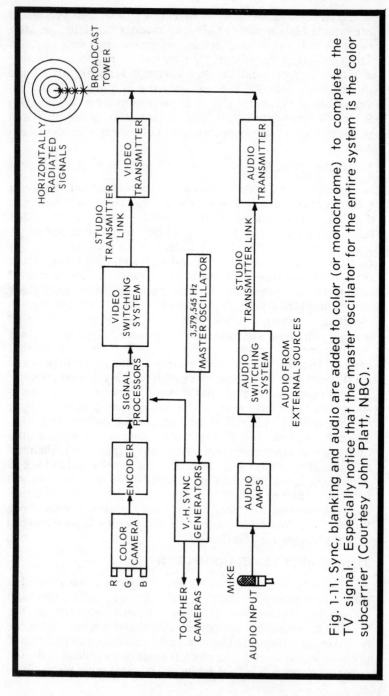

Fig. 1-11. Sync, blanking and audio are added to color (or monochrome) to complete the TV signal. Especially notice that the master oscillator for the entire system is the color subcarrier (Courtesy John Platt, NBC).

the overall system so you will have a workable mental picture of the entire process. The additional necessities in either color or monochrome are the vertical and horizontal sync signals and sound.

Both camera(s) and the signal processing circuits must receive blanking **vertical, and horizontal sync pulses** for sweep-timing which controls camera operation. The same synchronizing pulses (Fig. 1-6) are transmitted over the air and used by any receiver for horizontal and vertical sync. In older sets, blanking pulses eliminated vertical retrace lines that would appear after each field, while horizontal blanking pulses blanked the CRT beam during horizontal retrace periods. Most color receivers now have multiple means of vertical and horizontal regenerated pulse blanking instead of the unsophisticated methods used especially in early monochrome TV. The sync pulses, of course, are of greater amplitude than the blanking pulses; they extend into the blacker than black region and so are easily separated by the receiver and diverted to the HV synchronizing circuits. Transmitter sync and blanking pulses are precise, crystal-controlled rectangular pulses that are binary counted down from a master oscillator to the exact microsecond duration and repetition rates required, and then combined as in Fig. 1-6 for broadcast.

The audio portion of the transmitted signal is produced by typical studio equipment (microphones, switching systems and amplifiers). Today, most such equipment is transistorized and the handling of this portion of each telecast is relatively simple except for the ever vigilant engineer "riding gain" on the sound mixers and amplitude controls. The output goes through the studio-transmitter link to the audio section of the transmitter where it is combined with that of the video transmitter. Both carriers are broadcast within the 6-MHz allowable bandwidth (Fig. 1-2). The sound is centered in the transmission envelope, at 5.75 MHz, 50 kHz wide, and exactly 4.5 MHz separated from the video carrier at 1.25 MHz. In any monochrome or color transmission, therefore, the sound is always 4.5 MHz **above** the video carrier and should so be displayed on any receiver tuner response curve over a relatively flat-top response. The 4.5 MHz "intercarrier" difference between the audio and video carriers is given greater attention when we discuss receiver sound in Chapter 6.

The transmitting antenna is usually placed on top of tall city buildings or equally high towers so that it may radiate a horizontally polarized signal that is basically circular,

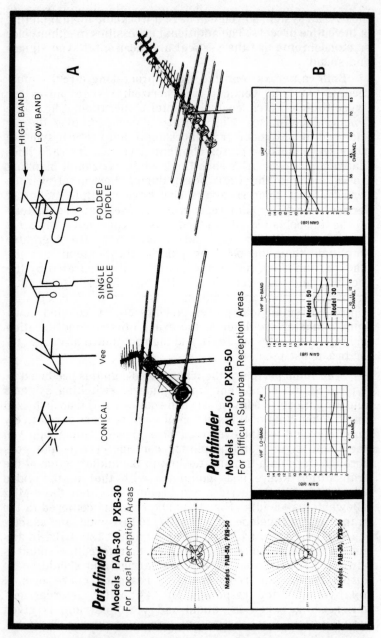

Fig. 1-12. Older types of VHF antennas (A). New Jerrold log periodic local or suburban antennas with separately directed front arrays (B).

although the broadcast pattern may be shaped. Both video and audio are radiated from the same structure through a bridge arrangement called a diplexer that permits the two transmitters to use the same antenna without disruptive interaction. When the bridge is balanced, there is no harmful feedback from one to the other and each transmitter may deliver full power to the single radiating antenna. The antenna and transmission lines must match closely, or delayed images can be reflected from the antenna back to the transmitter, reflected a second time, then radiated with a time delay that is twice the electrical length of the initial transmission line.

RECEIVING ANTENNAS

Because of earth-parallel polarization, receiving antennas must also accept electromagnetic television signals in the horizontal plane and be tuned to pick up either the VHF, UHF, or VHF-UHF combined transmissions. Naturally, there has to be a transmission line. And this line, like that of the transmitter's, has to be matched to the antenna and receiver to avoid what are known as standing waves (commonly referred to as the standing wave ratio or SWR), which induce ghosts and result in a loss of antenna gain, thus limiting the signal input at the RF terminals of the receiver.

With more than 20 years behind us in television transmission and reception, manufacturers have come a long way from the conical, vee, single folded dipole narrow hand antennas (Fig. 1-12A) to very complex arrays that are multituned, computer designed, often UHF-VHF combinations, and some, like the new Jerrold Pathfinder log periodic local and suburban antennas in Fig. 1-12B, have directional arrays for separate UHF station selection. Of course, both antennas receive full VHF-UHF transmissions. It's interesting to note the difference in gains as receiving elements are increased in number for suburban reception.

With a UHF station at, say Channel 45, we come down a 300-ohm oval shielded or non-shielded line (depending on reflections or noise) at 11 db of gain into the RF terminals of the receiver. Or, should the antenna be a 75-ohm variety (as this one can be also) a matching coaxial cable (single conductor) lead-in could be used that would connect directly into the tuner of the better new receivers. With the 300-ohm cable, the signals pass through a balanced-to-unbalanced transformer-like termination (balun) to match the 300-ohm line impedance to the 72-ohm receiver.

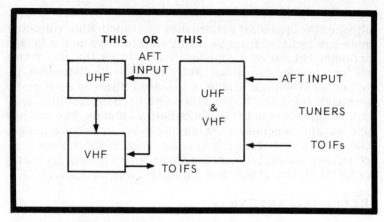

Fig. 1-13. Separate or combined UHF-VHF tuners are found in the newest television receivers. Most have AFT oscillator control from the third video IF.

TUNERS

Modern VHF tuners are usually either switch or turret types, while UHF tuners are virtually all the rotary capacitor variety—at least for the time being (Fig. 1-13). Varactor-controlled tuners, however, are rapidly coming more into use, although there are a few small problems with the control diodes yet to be solved, and the DC energizing voltages usually have to be well regulated. Some receivers also offer push-button controlled UHF selectors, and others have 13 or more detented channels so that the rotors switch into place. In any event, contemporary UHF tuners usually have a simple oscillator and mixing diode, without an RF amplifier, although there may be UHF RFs very shortly. The VHF counterparts all have at least one RF amplifier, a mixer, and a local oscillator, called by many in the past, the first detector.

Individual 6-MHz bandpass RF signals from the 12 VHF channels and 70 UHF stations may then be selected by the receiver's tuners, amplified, and converted to an intermediate IF frequency of 44.25 MHz. The UHF tuner, normally having no RF amplifier, picks up additional gain from the VHF tuner with added amplification by the VHF RF amplifier and mixer. The VHF oscillator is inoperative (usually no B+) while ultra high frequencies are being received. VHF-UHF channels and the corresponding frequencies are listed in Table 1-1.

IF & AFT AMPLIFIERS

In some receivers, coupling, certain bandpass shaping, impedance matching, and selectivity is insured by a tuned circuit between the tuners and video IF amplifiers. While in others, nothing more than a coaxial cable links the RF section with the IFs. In all instances where there are links, they have to be sweep aligned to match the tuner with the IF if either radio frequency or intermediate frequency circuits are replaced or modified. Where repairs simply mean physically replacing a same-value resistor or capacitor, alignment is usually not necessary unless, of course, the component was defective when the initial alignment was executed.

IFs are the basic bandshaping medium-to-high gain amplifiers that are normally stagger tuned so the overall—not individual circuit—response determines the bandwidth of the circuit. The first video IF is almost always controlled by a varying DC voltage derived after the video detector, called automatic gain control (AGC). The second stage is often regulated by the same voltage too, so that the IF amplifiers will not overload and cause sync problems, usually vertical. Traps to remove unwanted carriers, such as audio, are also included, as are various tuning (coupling) schemes to get signals from one stage to the next with maximum linearity, least loss, and best gain. Signals coming into an IF strip are at best in the millivolt range, and must be amplified to 4 or 5 volts, with load, before being detected and passed on to the video amplifiers. The critical stage in any IF amplifier is usually the third IF, since it passes the most current for the greatest voltage swing and IF amplification.

As you observe in Fig. 1-14, tube-type video IFs usually have at least two stages (sometimes three) but three or four transistors are always used in this same circuit because of the smaller voltage swing with the 22 to 35-volt DC supplies. Vacuum tube IFs often have 180-volt static sources, and "stacked," can function at between 200 to 300 WVDC. However, the IF vacuum tube stage doesn't have the small signal transistor's bandpass, longevity, easy DC operation, lack of harmonics, and little or no case dissipation. A tube always runs hot and, in many instances, has forever been a demon to tube sockets and printed circuit boards.

The AFT automatic fine tuning system is best described as a frequency discriminator to DC converter that should linearly feed back a proportional DC correction voltage for a deviation on either side of the 45.75-MHz video IF carrier. Such action controls the local oscillator in the tuner and puts the

Channel No.	Freq. In MHz	Channel No.	Freq. In MHz
2	54-60	43	644-650
3	60-66	44	650-656
4	66-72	45	656-662
5	76-82	46	662-668
6	82-88	47	668-674
7	174-180	48	674-680
8	180-186	49	680-686
9	186-192	50	686-692
10	192-198	51	692-698
11	198-204	52	698-704
12	204-210	53	704-710
13	210-216	54	710-716
Very High Frequencies (VHF)		55	716-722
14	470-476	56	722-728
15	476-482	57	728-734
16	482-488	58	734-740
17	488-494	59	740-746
18	494-500	60	746-752
19	500-506	61	752-758
20	506-512	62	758-764
21	512-518	63	764-770
22	518-524	64	770-776
23	524-530	65	776-782
Ultra High Frequencies (UHF)			

Ultra High Frequencies (UHF)

Channel No.	Freq. In MHz	Channel No.	Freq. In MHz
24	530-536	66	782-788
25	536-542	67	788-794
26	542-548	68	794-800
27	548-554	69	800-806
28	554-560	70	806-812
29	560-566	71	812-818
30	566-572	72	818-824
31	572-578	73	824-830
32	578-584	74	830-836
33	584-590	75	836-842
34	590-596	76	842-848
35	596-602	77	848-854
36	602-608	78	854-860
37	608-614	79	860-866
38	614-620	80	866-872
39	620-626	81	872-878
40	626-632	82	878-884
41	632-638	83	884-890
42	638-644		

Table 1-1. VHF and UHF channel designations.

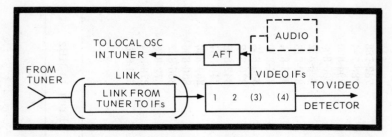

Fig. 1-14. Block diagram of the tuner and video IF link connector on many older sets, using 2, 3 or 4 stages of video IF. Automatic fine tuning (AFT) keeps the local oscillator in the receiver from drifting off proper tuning.

tuner back on the selected channel frequency. Almost universally, AFT correction voltages are taken off the collector of the third IF, similar to the method used to pick off audio signal.

VIDEO DETECTOR & AMPLIFIERS

The video detector (sometimes called the second detector) is usually a half-wave rectifier diode (Fig. 1-15), although there are probably full-wave IC rectifiers on the way. Normally, it receives the IF signals on the cathode, blocking the positive portions of diode conduction but passing negative composite video so that sync pulses and noise will be pointed toward blacker than black and, therefore, not appear in the picture. Usually, in color anyway, a video detector has a 4.5-MHz intercarrier sound trap before it as a last resort to block sound frequencies before they enter the amplified video chain

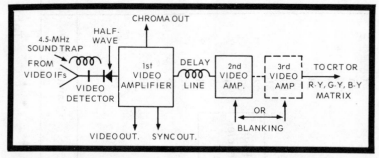

Fig. 1-15. Block diagram of a diode half-wave video detector followed by two or more video amplifiers with an 0.8-microsecond delay line in between.

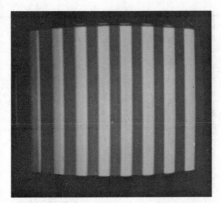

Fig. 14-55. Red demodulator output open.

Fig. 14-58. Good rainbow color-bar pattern. The first yellow-orange bar is usually hidden.

Fig. 14-56. Green demodulator output open.

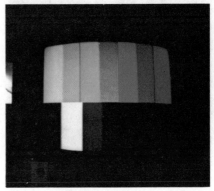

Fig. 14-59. Compare this fully saturated NTSC pattern with the gated rainbow in Fig. 14-58.

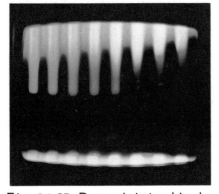

Fig. 14-57. Demodulator blanking failures are difficult to handle with a partial open or short

Fig. 14-60. Blue demodulator cutoff.

33

Fig. 14-61. Open input to the red demodulator.

Fig. 14-64. A bypassed 3.58-MHz filter shifts the phase of the subcarrier oscillator.

Fig. 14-62. Shorted green demodulator input.

Fig. 14-65. If the hue corrector bias resistor decreases in value, green disappears on the right.

Fig. 14-63. Blue demodulator input failure.

Fig. 14-66. Another 3.58-MHz filter develops added capacity and causes a phase shift.

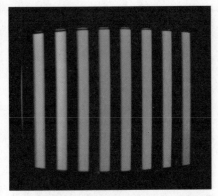

Fig. 14-67. Subcarrier oscillator output phase shifts left with a partial AC short.

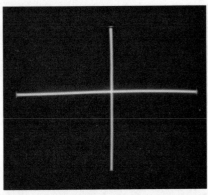

Fig. 14-70. This generator pattern has single crossbar that can help in static convergence.

Fig. 14-68. This color-bar is useful for vectorscope analysis.

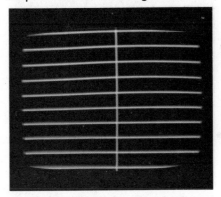

Fig. 14-71. Ten lines of horizontal bars and one vertical. Observe blue fringing top and bottom.

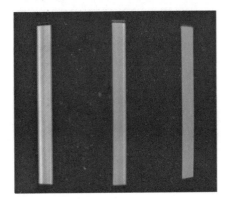

Fig. 14-69. The presence of R-Y, B-Y and G-Y permits a quick-check of the chroma output.

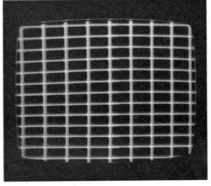

Fig. 14-72. Crosshatch pattern from the first generator used has undesirably thick lines.

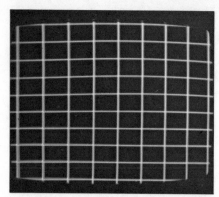

Fig. 14-73. A thinner and more uniform crosshatch shows misconvergence more plainly.

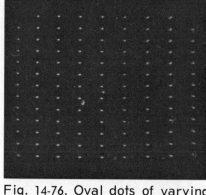

Fig. 14-76. Oval dots of varying size can cause convergence confusion.

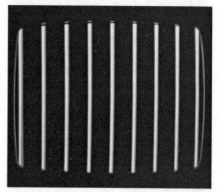

Fig. 14-74. These vertical bars look good, but are much too wide for best convergence.

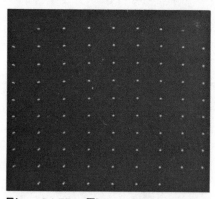

Fig. 14-77. Tiny, symmetrical dots readily show blue and red misconvergence at either edge.

Fig. 14-75. These horizontal bars are almost useless because of line width variation.

Fig. 14-78. The brand-new Telequipment D66 is an excellent troubleshooting scope.

and become interference in the picture. Around the video detector there are also peaking coils to increase high-frequency response.

The first video stage is the initial amplifier of demodulated composite video containing all sync, monochrome, and chroma information to be used by the remainder of the receiver's circuits. The capacitance, inductance, and passband characteristics of the video detector and video IFs often cause a video "tilt" that usually requires compensation in the chroma IFs. Also, because of the additional signal processing in the chroma circuits, there has to be an 0.8-microsecond delay in the luminance channel so that chroma and luminance information can drive the final picture tube amplifiers or CRT grids at the same instant in time. The first video amplifier, therefore, supplies some of the usual video, and certainly chroma and sync information, to the various other processing circuits in the receiver.

After the delay line, the luminance (video) signals are sent to one or more additional amplifiers, depending on the requirements of the cathode ray tube and the manufacturer's design team. These add further gain to the luminance monochrome signal which is usually no more than 3.6 MHz on good color receivers, equal to or less than 2.8 MHz on poor ones, and somewhere in between when processed by monochrome receivers, depending more or less on price. If the bandwidth on a color receiver is too wide, there will be a collision between luminance and chroma, and this will show as fine-line (or herringbone) interference all over the screen. We'll look at receiver frequency response later with the vertical interval test signals (VIT) that are dependable evaluators of color receivers, with but few predictable reservations.

AGC, SYNC & NOISE CIRCUITS

AGC and noise circuits vary from receiver to receiver, but sync separators and amplifiers are relatively standard. The ideal circuit would be something like the diagram in Fig. 1-16, where the noise gate would act on both the AGC amplifier and sync separator alike when noise pulses arrive. If it did, here's what would happen: Incoming negative video would supply both the AGC amplifier and sync separator with picture and sync information that each could use for its respective operations. The automatic gain control amplifier is AC operated from flyback pulses obtained through a winding on the horizontal output transformer. The sync separator must

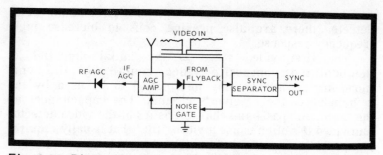

Fig. 1-16. Block diagram of the AGC, sync separator and noise gate combination found in some receivers.

strip all video from the composite video-sync signal and present only horizontal and vertical sync information to the receiver's sync circuits. As more video signal is received, the AGC amplifier conducts harder and supplies more forward bias AGC to the (semiconductor) video IF amplifiers which begin to shut off conduction current when forward DC becomes abnormal. On very strong signals the RF AGC delay begins operating when the IF AGC forward voltage has risen more than 0.7 volt (the forward diode drop) and the RF amplifier in the tuner amplifies less. Now, the IF transistors and the RF transistor or FET all conduct less, resulting in greatly reduced overall gain. In vacuum tube receivers, the bias would be negative so that tube grids would be driven less positive, thereby reducing conduction by cutting down the flow of electrons through the tube.

As you must already know, when there are no sync pulses, the horizontal and vertical multivibrators operate all by themselves at set time constants and DC levels and, therefore, produce a steady raster. With a signal input, however, the sweep circuits must be timed so they are in step with the transmitted signal in all respects. The incoming video signal at the sync amplifier input is in negative polarity. So if a large spike of transient voltage (any unusual inside or outside interference—usually outside) spears the composite video, the noise gate can be designed so that it shunts this transient to ground, and at the same time biases the AGC. Then, both the sync and video signals are blanked, the spike of errant voltage (only a few microseconds or so in duration) is shunted aside, and an observer never knows the difference. As you can see, noise circuits, designed properly, are thoroughly useful in any set, especially the better color receivers.

With respect to video transmission, negative video modulation is used in this country for better automatic gain control, since sync tips are held constant. Also, the peak transmitter power output is 30 percent greater with negative modulation than with positive modulation. With negative modulation, noise also appears as **black** interference instead of white, a condition that is certainly less noticeable to the eye.

VERTICAL OSCILLATOR & OUTPUT

These circuits are mostly the feedback type, with an extra driver included in the transistorized versions (Fig. 1-17). Really, all they amount to is a pair or tubes (probably in the same envelope) three or more transistors or, in one instance, a single integrated circuit mounted on a heat sink. In all cases, however, feedback signals from the output tube, or driver transistor, go to the oscillator to sustain "flywheel" oscillations during times of no incoming sync signals. There are linearity controls to correct conduction curves and height controls to stretch the picture to cover the face of the cathode ray tube. Sync pulses at a 59.94-Hz (16.66 msec) rate are differentiated and used to sync the vertical oscillator which must have an accuracy of 3 microseconds, or one part in 5,000—a tolerance that is 8 times more critical than the horizontal scan time. Therefore, sync troubles will usually affect the vertical oscillator first, if sync jitter or other instability is the problem.

There's nothing especially new in vertical oscillators, but the output sections of some transistorized modular receivers have changed drastically, and in several of these, there is no longer any vertical output transformer as we've known it since the advent of television—only a complementary push-pull arrangement of two stacked transistors with a 500-mfd coupling capacitor feeding to the deflection yoke. This circuit

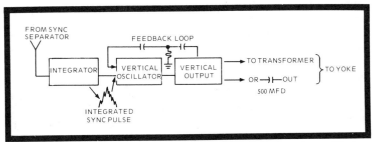

Fig. 1-17. Block diagram of a typical vertical oscillator and output sweep system. It is driven by an integrator drive pulse and coupled by transformer or capacitor to the vertical deflection yoke, with further coupling to convergence and pincushion subsystems.

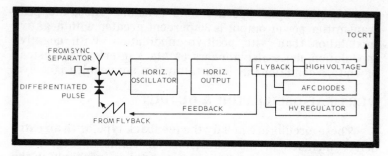

Fig. 1-18. Horizontal sweep system block diagram including oscillator, output, high-voltage, regulation and feedback.

is a logical development from a similar audio speaker-driver circuit first introduced, we believe, by Motorola in 1969 or 1970, depending on what is considered the model year.

The total vertical trace time is 16,662 microseconds (16.662 milliseconds), a blanking period of about 1.4 millisecond for each field. Two fields constitute a frame, and 30 frames per second is just above the 24-frame flicker rate our eyes recognize. The input circuit to all vertical oscillators is an integrator that is actually a low-pass filter designed to screen out the faster horizontal pulses and deliver only slowly integrated vertical pulses that either cut off or drive on the vertical oscillator, depending on the particular sync system.

HORIZONTAL OSCILLATOR & OUTPUT

The horizontal oscillator, like the vertical oscillator, has sync and feedback circuits, too, but they are somewhat different (Fig. 1-18). The **differentiated** horizontal sync pulses are the timing chain for this circuit when the set is tuned to a signal. Its input is, of course, an RC differentiator, with the network resistor connected to ground. The resistor and its associated capacitor form a time constant that will both block the vertical sync frequencies and allow a reasonable rise and fall time for timing the horizontal oscillator. There are, however, automatic frequency control diodes (AFC) that take the output transformer's feedback pulse, compare it with the incoming sync signal, and then put out a DC correction voltage for oscillator lock. If the oscillator free runs correctly with no incoming signal, then stays locked in sync with incoming signal, you know that both sync and oscillator circuits are operating as they were designed. An out-of-sync condition with no incoming signal denotes oscillator problems, while a sync clip with incoming signal tells you there is probably sync compression somewhere or other faults in, or ahead of, the sync separator.

A semi-trapezoidal or rectangular-shaped pulse signal is produced by tube or semiconductor oscillators and is used to drive the horizontal output stage. The output stage supplies a horizontal output transformer, horizontal deflection yoke, pincushion transformer (on most sets), and the dynamic convergence coils. This takes a sizable power tube or one or two heavily rated transistors or silicon controlled rectifiers to drive such circuitry. In combination, the oscillator and horizontal output are among the most critical parts of a color television receiver. For, in addition to exciting the various inductors that deflect and converge the three CRT beams, the collapsing magnetic field of the horizontal output transformer must form a high-voltage spike that can be converted into focus voltage, B-boost voltage, other pulse drive sources, and 25 to 30 kilovolts to light the picture tube. Also, all this has to have regulation so that the CRT intensity will remain relatively constant and the picture tube will not draw too much beam current during maximum picture brightness. This is why the high voltage is always adjusted with the brightness control turned to **minimum**, since you don't have to depend on the regulators which are not fully operating at this point.

THE CHROMA CIRCUITS

The last of the video receiving blocks in this initial description are the chroma circuits (Fig. 1-19). If you will simply consider these color processors as just another subsystem among the previous groupings and remember each of the basic subsections, you should have little difficulty.

Initially, you recall that the chroma information out of the transmitter contains double and single sideband modulation on either side of the 3.58-MHz suppressed carrier. The chroma signals are interleaved among the monochrome modulation at sum and difference multiples of the horizontal scanning frequency. The receiver actually accepts I and Q signals as transmitted, but the set is both a little less broadbanded and also shifts the phase of the incoming chroma 33 degrees right (clockwise) so that R-Y and B-Y information is processed instead. And since the eye can't tell without immediate reference that I information above 0.5 MHz is not included, the receiver also attenuates the 0.6 to 1.5 MHz color information, permitting just the double 0 to 0.5 MHz I and Q sidebands in the form of R-Y and B-Y to pass to the color amplifiers.

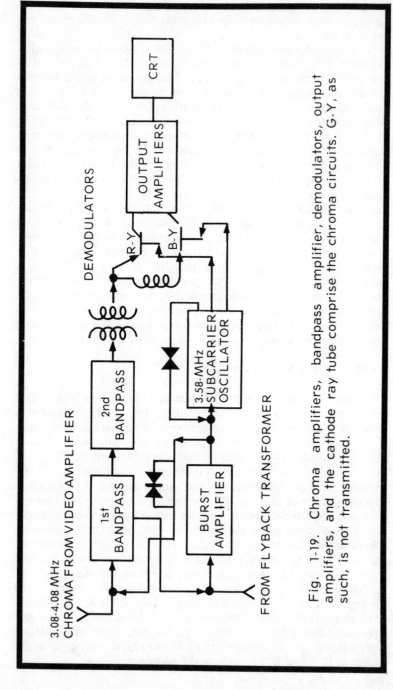

Fig. 1-19. Chroma amplifiers, bandpass amplifier, demodulators, output amplifiers, and the cathode ray tube comprise the chroma circuits. G-Y, as such, is not transmitted.

The color amplifiers are called **bandpass** amplifiers because their function is to conduct and amplify only the 3 to 4 MHz information 0.5 MHz on either side of the 3.58-MHz subcarrier. Other chroma circuit terminology includes the chroma IFs and sometimes, in integrated circuits, gain-controlled amplifiers. But they will still do the same thing: permit passage of only color frequencies between 3.08 and 4.08 MHz for eventual processing by the color demodulators. Of course, there are DC correction and control circuits along the way, such as automatic chroma control and, in one receiver, an additional DC automatic control of the second chroma amplifier.

Also recall the 8-11 cycles of burst on the back porch of the horizontal sync pedestal. This is the color sync signal which is applied to a burst amplifier that is gated into conduction by a winding on the flyback transformer every 63.5 microseconds, or at the end of each horizontal scan line. Color sync rings a crystal or is compared with feedback from the 3.579,545-MHz receiver-generated subcarrier oscillator. If the oscillator frequency differs, a DC correction voltage is produced that syncs the local oscillator with that of the suppressed carriers at the transmitter. The amplitude of the incoming burst is somewhat proportional to the received chroma, and another DC voltage is generated to control conduction of the first chroma (bandpass) amplifier within relatively small excursions.

The regenerated chroma subcarrier now supplies one or more signals to a pair of diodes, a pair of transistors, three transistors, or an integrated circuit known as the chroma demodulator. The conducting demodulators pass the chroma signals to either an output matrix comprised of additional transistors, or matrix the chroma internally, or supply R-Y, B-Y, G-Y information to the grids of the cathode ray tube where the chroma and luminance signals are matrixed. This latter method, of course, is the older system and less and less manufacturers are using it since there are many advantages to passing red, blue, and green information directly to the cathodes of the picture tube already mixed with brightness intelligence. You'll see why when we analyze the several types of demodulators.

SOUND SECTION

The sound section (Fig. 1-20) completes this preliminary discussion of the receiver. The difference signal between the 45.75-MHz video carrier and the 41.25-MHz sound carrier creates a new frequency modulation carrier at 4.5 MHz. The

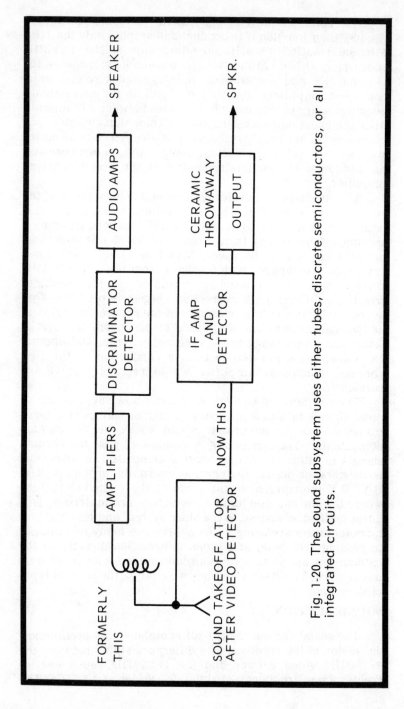

Fig. 1-20. The sound subsystem uses either tubes, discrete semiconductors, or all integrated circuits.

SPEAKER

AUDIO AMPS

DISCRIMINATOR DETECTOR

AMPLIFIERS

FORMERLY THIS

SOUND TAKEOFF AT OR AFTER VIDEO DETECTOR

NOW THIS

IF AMP AND DETECTOR

CERAMIC THROWAWAY

OUTPUT

SPKR.

original sound carrier is trapped out. The difference signal must pass through a frequency discrimination circuit, be amplified, and then detected by various means that have included several types of discriminators, ratio detectors, quadrature locked oscillators, gated beam detectors, etc. The latest systems use integrated circuits, with various internal amplifiers, peak detectors, and output amplifiers all rolled into one. The newest audio output circuits have complementary transistors like the vertical circuit previously described and an AC coupling capacitor instead of a DC operated transformer. Also, some audio circuits, and some video and sync circuits too, are being made of thick film and transistor chips on ceramic substrates. All will be what the industry calls "throwaways"; in other words, non-repairable.

FM detection converts carrier frequency deviations to a current response, where the current generates a ratio-like voltage that is proportional in duration and amplitude to any and all carrier variations.

SOLID-STATE RECEIVER

To conclude this first chapter, it seems the discussion is completely up to date with the block diagram of a virtually all-integrated circuit color television receiver (Fig. 1-21) developed by Motorola, Phoenix, Ariz.

The tuner can be a printed circuit, varactor-operated type; the IFs, AFT and video detector, a trio of ICs; the AGC, noise and sync are in an MC1345 IC, the horizontal AFC and oscillator, an experimental XC1391; tuner regulator, an MVS460; the power supply, an MFC4060; sound IF and audio outputs, another pair of ICs; the chroma demodulator, an MC1326; chroma processor, an MC1398; and the vertical deflection, a second experimental XC1390—a circuit that won the Broadcast and TV Receiver IEEE Group's 1971 award. The day of an all IC receiver may not be long in coming.

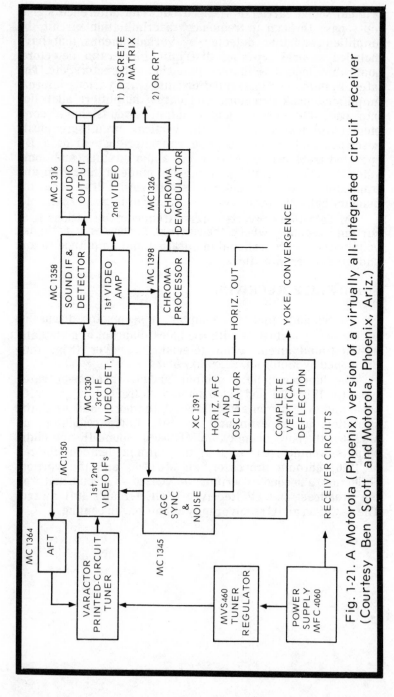

Fig. 1-21. A Motorola (Phoenix) version of a virtually all-integrated circuit receiver (Courtesy Ben Scott and Motorola, Phoenix, Ariz.)

QUESTIONS

1. What other energy travels at the same speed as light?
2. Light and what travel at the rate of 3×10^8 meters per second?
3. Radio waves consist of what kind of fields?
4. Modulated transmitted waves must undergo what two variations?
5. Define amplitude (AM) and frequency (FM) modulation?
6. What sidebands are especially important in the study of color television?
7. Name two sidebands on either side of the color subcarrier.
8. How many TV channels are there presently allocated and what is the individual bandwidth?
9. What is the VHF-FM bandspread? UHF bandspread?
10. How do I and Q sidebands relate to burst?
11. What are the differences in degrees between I, Q, and burst?
12. What are the time intervals for vertical and horizontal blanking?
13. Define the horizontal and vertical blanking intervals in terms of time.
14. Sync information is broadcast only during _____?
15. Black is what percentage of the composite waveform's amplitude? White? Sync Peaks?
16. What are the vertical and horizontal repetition rates for color TV transmissions? Chroma subcarriers?
17. Chroma is inserted at what basic frequency on either side of the subcarrier?
18. What colors comprise the luminance signal and what are their proportions?
19. Draw a diagram of the transmitted I, Y, and Q signals from camera to transmitter.
20. In transmission, is audio above or below the video carrier, and by how much?
21. Why must antenna and transmission lines at the transmitter and receiver be matched?
22. What is the sync reference for the entire color TV transmitter?
23. What happens when receiver video IF amplifiers are allowed to overload? Will audio sometimes buzz?
24. Describe an automatic fine tuning circuit.
25. Do we have positive or negative video modulation in the U.S., and why?

Chapter 2

Colorimetry & Picture Tubes

Color is a combination of light and dark, tints and broad displays that can be described as **brightness, hue**, and **saturation**, where **brightness** (luminance) is identified as the various shadings (usually 10) from very light to very dark; **hue** describes the actual color such as red, blue, and green; and **saturation** tells whether these colors are intense (fully saturated) or have considerable white content and are, therefore, pastels.

COLOR MIXES

An artist mixes paints such as yellow and blue to produce green, red and yellow for orange, and blue and red to make purple. His primary colors are blue, red, and yellow. And, in mixing these individual pigments, he finds they **absorb** particular hues; therefore, it is a process identified as a **subtractive** color mixing.

Color television, however, uses an **additive** method where light (Y for luminance) is added directly to primary reds, blues, and greens, so that R(red)-Y(luminance, B(blue)-Y(luminance), and G(green)-Y(luminance) are the color signals and in the receiver are either combined with luminance in the output color amplifiers for passage to the cathodes of the picture tube (in the late model receivers), or continue as separate luminance and color signals to the grids and screens of the cathode ray tube where they are matrixed internally. This luminance information contains not only brightness, but also includes all the fine detail you see in any monochrome or color picture because color fills in only the larger areas of any televised scene.

As indicated above, the color mixes in color TV are entirely different from those made with pigments, and this is shown graphically in Fig. 2-1. Blue and green constitute cyan; blue and red make magenta; red and green produce yellow; and red, blue, and green combine in certain proportions to yield white. On the other hand, when any primary hue is subtracted from white, the result is a complementary color such as yellow being the complement of blue, with pairs cyan and red, green and magenta constituting equivalent complements, too. Other intermediate colors result from changing the proportions of the primaries. For instance, a certain mix of considerable red and green forms orange; chartreuse also comes from measured quantities of mostly green and some red, etc.

Completely saturated colors seldom appear in nature's settings. Even the sky and sea are somewhat less blue and green than full saturation might require. Consequently, the majority of colors you see are unsaturated and this means that most of the 1,000,000-odd phosphors on the face of the cathode ray tube are usually lighted to some extent, depending on the sweep and intensity (amplitude) of the incoming chroma signals.

COLOR FREQUENCIES & APPLICATIONS

Believe it or not, the eye can recognize more than 30,000 colors. These colors appear in the radiant energy spectrum between 400 and 700 nanometers (formerly millimicrons) from lower frequency ultraviolet progressively to blue, green, yellow, orange, and red, the upper visible frequency (Fig. 2-2). In round numbers, you see light in the visible spectrum at

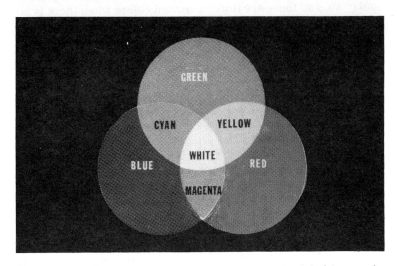

Fig. 2-1. Primary colors red, green, and blue mix to form cyan, white, magenta, yellow.

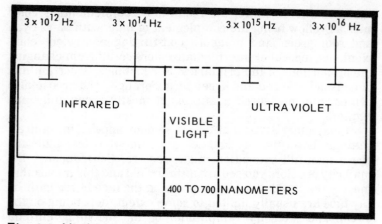

Fig. 2-2. Visible light appears in both the radiant energy spectrum and frequency range between nonvisible infrared and ultraviolet.

frequencies between 3×10^{14} and 3×10^{15} Hz, just following infrared and before ultraviolet.

The primary television colors, red, blue, and green, have spectrum wavelengths of 680, 470, and 540 nanometers, respectively, and have been deliberately selected to deliver the greatest range of suitable colors. For instance, in Fig. 2-1, 50 percent of two of the three colors produce magenta, yellow, and cyan, and these are fully saturated colors when there is no third hue to intervene. When all three saturated colors combine (unmodulated), the result is (or should be) a uniform gray-white raster. At this point, all phosphors are lighted with the same intensity and, to the eye, appear to have equal energy or brightness.

Experimentally, we find that a reference white light with the three TV primary colors of the same intensity will produce apparent brightnesses of 59 percent green, 30 percent red, and 11 percent blue. In the real life picture tube, of course, these percentages are not valid because of the different efficiencies of the RGB phosphors which, incidentally, now favor green as the weaker (or reference) color, and red and blue as the stronger colors. Only a few years ago, red was decidedly the least efficient. But, with new Yttrium red phosphors, this color has gained considerable comparative strength.

For reference and general information, the chromaticity diagram (Fig. 2-3) from Sylvania Color TV Clinic Manual I shows you how the NTSC color triangle is put together, and where the various colors lie throughout the curve. The numerals surrounding the X-Y plot are the wavelengths in nanometers of the different visible colors.

NTSC & SIDELOCK COLOR GENERATION

The highly useful but usually more expensive NTSC generator delivers carrier, sync, and pulse-shaped voltage levels to the chroma circuits of any receiver (Fig. 2-4), resulting in fully saturated colors that may number between 3 and 6, depending on the design of the individual generator. The amplitudes of the rectangular pulses in the set can be both positive and negative, depending on the action of the receiver's chroma demodulators, and each pulse occupies a unique level and position. The position and level tells you which red, blue or green output you're viewing, the condition of its demodulation, and the efficiency of its RGB-(Y) amplifier—all excellent things to know when either designing and-or troubleshooting chroma processing circuits. Notice the

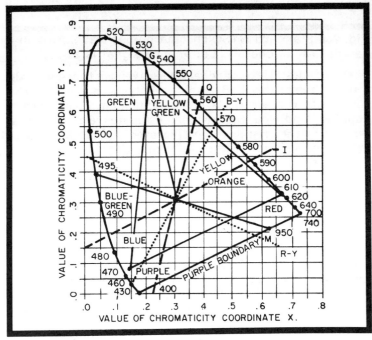

Fig. 2-3. Sylvania's color television chromaticity X-Y diagram with colors identified in the radiant energy spectrum in terms of nanometers.

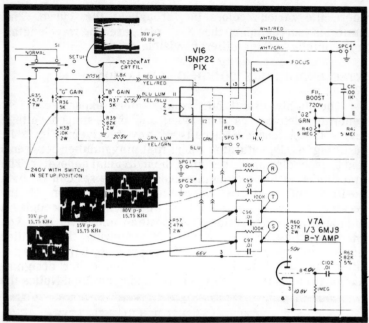

Fig. 2-4. Standard NTSC chroma output waveforms in a Zenith CM-114 chassis. Luminance signal is at top.

different pedestal levels and relative waveform time-intervals of the pulses at the RGB amplifier outputs in Fig. 2-4. The pulse peaks represent the largest amplitude swings out of each amplifier, and you can roughly estimate relative timing (without a triggered sweep oscilloscope) by the position of negative-going 11-microsecond blanking pulse before the start of each new cycle.

A little less than two cycles of each waveform are showing in each of the RGB-Y waveforms in Fig. 2-4. For instance, the B-Y amplifier swings fully positive at the beginning of every new cycle, then begins to descend while the R-Y amplifier output is rising. Meanwhile, the G-Y amplifier swings almost fully negative several microseconds after blanking. This sequence should indicate that the first color bar would be blue, the second magenta, and the third might well be a combination of red-blue (the total absence of a time base is confusing) which could equal a third color, with blue substituted for yellow-white. So red-blue equals red-yellow + white equals 0 and yellow equals red + white. Therefore, yellow is the third color in the sequence, at least for the present best estimate. Shortly, we'll demonstrate what you see exactly in a gated (sidelock) rainbow generated signal.

An NTSC color-bar generator (Fig. 2-5) produces a constant frequency of 3.579,545 MHz, the same as the chroma subcarrier oscillator in any color receiver, and produces saturated colors in the receiver circuits by both gated pulse amplitudes and certain phases that are time-delayed by taps on a fixed or variable delay line. Such generators, for instance, produce red at 76.5 degrees, blue at 192 degrees, and yellow at 12 degrees in any normal 4-quadrant vector diagram.

The sidelock (or gated rainbow) generator is a somewhat less expensive instrument, depending on quality, overall performance and circuit composition. The term sidelock refers to the fact that the output sine-wave generator signal operates at exactly 15,734 Hz, minus the fixed color oscillator frequency of 3,579,545 Hz. This generator, then, produces one less alternation in each 63.5-microsecond horizontal scan

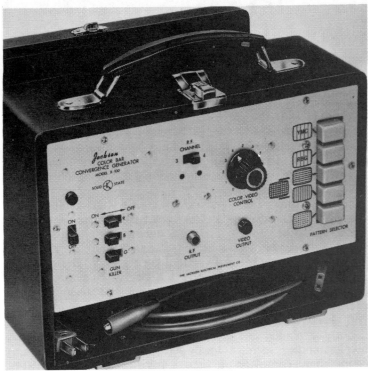

Fig. 2-5. Jackson X-100 NTSC type 3-bar and complement color generator with 315-kHz master oscillator, featuring RF and video outputs.

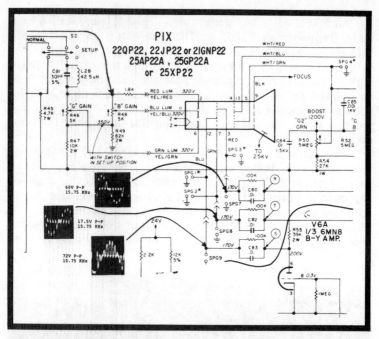

Fig. 2-6. Zenith CM-114S CRT input circuits with sidelock (rainbow) generated color output waveforms.

interval, but at the start and finish of this period it is completely in phase and the set doesn't know the difference. Therefore, color lock is practically possible when a color generator is offset by an exact multiple of the horizontal scanning rate, and the term "sidelock" is often used to describe such operation.

The original color generators simply had a blending of reds, blues, and greens, in that order, to produce a constant phase variation of from 0 degrees around the circle and back to 360 degrees. But with a bit more complex circuitry, such as countdown horizontal and vertical sync gating, and a master 190-kHz oscillator, it was found that twelve color bars should be produced that would cause the receiver to show variations of from yellow-orange, through reds, magentas, blues, cyans, and greens and some other shades in between. However, the generator blanks one of these color bars, and the receiver blanks another, so you actually see 9 or 10 color bars (one may be lost in overscan) on the face of any satisfactorily operating color receiver.

The diagram in Fig. 2-6 shows a later model Zenith (with virtually the same color output circuitry) excited by a gated

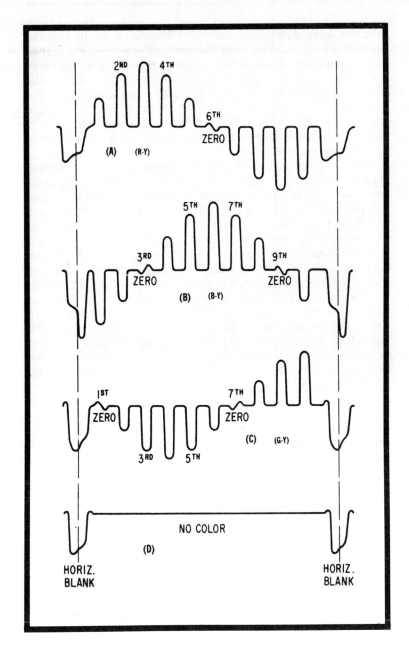

Fig. 2-7. R-Y, B-Y and G-Y reference patterns generated by a rainbow-type generator. Notice the zero nulls in all three patterns.

rainbow generator. The one problem here is that these waveforms were taken with the hue (or tint) control turned just enough to disturb the classic relationship for zero reference. Usually, the R-Y demodulated output nulls at the sixth bar, G-Y at the first and seventh bars, and B-Y at the third and ninth bars. Nonetheless, you can see the voltage levels quite clearly, and the illustration is a good example of virtually the same receiver circuits operating with NTSC-type and gated rainbow inputs.

For the record, let's include the classic tri-phase color-bar output as it should appear so you will have an accurate reference for future demodulated chroma evaluation and comparison (Fig. 2-7). The generator we might have used appears in Fig. 2-8. Other sidelock generators and color-bar plus vectorscope patterns are exhibited in the chroma and test equipment sections. Here, the vectorscope patterns are accurately timed with the gated rainbows to show exactly how and why these color bars are produced and where the RGB-(Y) receiver outputs combine to supply the various colors. A good receiver, we might add, does NOT blend various colors within the 10 color bars, but exhibits solid shades for each individual bar so they appear fully saturated. Pastels, wrong value, or faded colors usually mean there are defects in the chroma circuitry or, in the worst instance we can think of, you're looking at a poorly designed color receiver. And then, of course, any normal repair is a little more than difficult. As you will discover later in the vectorscope timing diagrams, however, only the R-Y and B-Y outputs are used, since these are Lissajous patterns, and only inputs 90 degrees out of phase can be used. If these exciting voltages were in phase, you'd have nothing but a right or left-slanted diagonal line.

WAVEFORM ANALYSIS

Just to keep the colorimetry in perspective, let's analyze a few points in these color test patterns and compare the generated colors in Fig. 2-9A with the idealized output drawings in Fig. 2-7 and the vectorscope color wheel in Fig. 2-9B. The first negative-going voltage is simply horizontal blanking that always precedes the start of the gated rainbow color pattern. In the first bar, with green at zero, two increments of negative-going B-Y and one positive R-Y produce a yellow orange color bar on the left. In time 2, two units of R-Y and one unit each of B-Y and G-Y (both negative-going) produce a full orange. The third bar, with two units of negative green, no B-Y, and a full three units of positive R-Y, the pic-

56

ture shows (pinkish) red. In the fourth bar, two units of positive R-Y, one unit of positive B-Y, and three units of negative G-Y resolve what is called magenta.

Notice that up until this time, there is virtually no greenish effect, and it is only when blue B-Y begins to go positive that the blue bars really enter the picture. When B-Y diminishes and R-Y goes negative, G-Y excitation rises positively and the final two bars are bluish green and green, respectively. After this follows the second cycle of blanking and the color-bar pattern begins all over again.

Even though we can make a logical case of the position of the green color bar at certain phases of the line frequency scan, it isn't really necessary. Remember, in Chapter 1, the original I and Q information contained green in both instances. In many receivers, even today, green is the negative output mix of the R-Y and B-Y amplifiers. Consequently, it's nice to

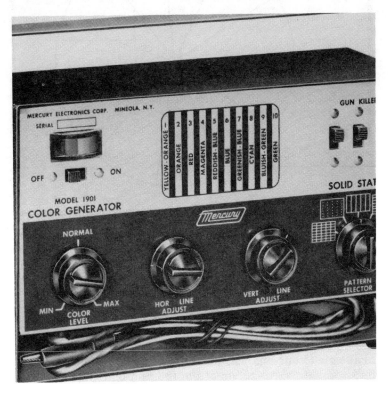

Fig. 2-8. Mercury's 1900-1901 inexpensive color-bar generator produces variable size lines and dots.

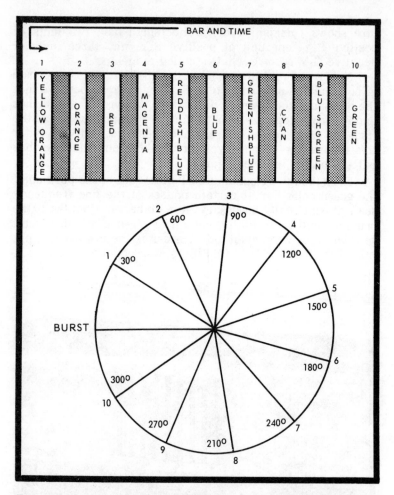

Fig. 2-9. A gated rainbow (sidelock) generator produces 12 color bars. Two are blanked, and either 9 or 10 should appear on any color CRT, one bar being lost, usually, in overscan (B). Vectorscope pattern of Figs. 2-7 and 2-9A combined. It can be seen on any vectorscope if the receiver is driven by a clean gated rainbow generator.

know the G-Y stage is functioning, but the result is really **academic** to our analysis since its effects are already taken into account by the action of the red and blue demodulators. Even in the newest receivers where green is demodulated separately from red and blue, the R-Y and B-Y demodulator-amplifiers (with luminance removed) still tell the gated rainbow color-bar story. New tri-phase demodulation, especially in integrated circuits with luminance directly added though, provides superior color since it delivers excellent greens, something that all X and Z or R-Y, B-Y systems don't always do.

USEFUL COLOR EQUATIONS

Here are some of the more useful equations that have been developed over the past 20-odd years to help explain our remarkable NTSC system of color:

$I = .74 (R-Y) - .27 (B-Y)$
$Q = .48 (R-Y) + .41 (B-Y)$ I and Q in terms of R-Y, B-Y

$I = .60R - .28G -.32B$
$Q = .21R - .52G +.31B$ I and Q in terms of red, blue, green

$Y = .30R + .59G + .11B$ Luminance in terms of red, blue, green

$R-Y = .96I + .63Q$
$B-Y = -1.11I + 1.72Q$ Color difference RGB-(Y) in terms
$G-Y = -.28I - .64Q$ of I, Q.

$R-Y = .70R - .59G - .11B$
$G-Y = .41G - .30R - .11B$ Color difference RGB-(Y) in
$B-Y = .89B - .59G - .30R$ terms of red, green, and blue.

There are also three basic oscillator equations that might help explain the origin of the sync and color subcarrier frequencies. Since separation of sound and video carriers at 4.5 MHz can be classed as an even 286th multiple of the line scanning frequency rate of 15,750, a new line frequency can be equated:

$$Lf = \frac{4.504,500 \times 10^6}{286} = 15,734.264 \text{ Hz}$$

or 63.5 microseconds

From this, a new vertical rate is determined, since each frame will consist of 525 lines:

$$Ff = \frac{15,734.264 \text{ Hz}}{525} = 59.94 \text{ Hz}$$

or 16,662 microseconds

And, finally, a 455 multiple is supplied to keep the subcarrier frequency a certain distance above the video carrier and still contain the I and Q sidebands. The subcarrier frequency now amounts to:

$$Sub_f = \frac{455 \times 15,734.262}{2} = 3,579,545 \text{ Hz}$$

or almost 282 nanoseconds

All equations after the first depend entirely on the line frequency as determined by the sound-video difference, a multiple and the original line scanning frequency. This is why a broadcasting station must hold its color subcarrier reference tolerance to 0.0003 percent, with a maximum rate of change of no more than 0.1 Hz per second, or 10 cycles in 3.6 million, a tolerance that must forever be crystal controlled at the transmitter and followed very closely by the receiver with its own crystal-controlled oscillator.

CATHODE RAY TUBES

In contrast to the screen surface in early tubes, current picture tubes are now beginning to or are already using a black background that surrounds the red, blue, and green phosphor dots to absorb light reflections. The black background also permits the use of higher transmission filterglass faceplates, and even allows beam overscan of individual dots, if this proves to be an advantage. There are also new red-emitting yttrium oxysulfide rare earth phosphors, and new green and blue emitting sulfide phosphors, too. To produce raster white light at 9300 degrees Kelvin, the anode current supplied by each of the three guns must equal 30 percent red, 31 percent blue, and 39 percent green, demonstrating that green is the least efficient phosphor and is now the reference color rather than red, which was always the weakest phosphor in the earlier days of color television.

Into the neck of each tube is fitted a precision triple-electron gun to limit, focus, modulate, and aim the three beams of electrons toward the same cluster of three-dot phosphors at any instant in time. Usually, the smaller the beams the better the focus. Static and dynamic magnets are positioned over each gun for controlled beam landing to make sure each beam hits the correct dot, plus a lateral blue magnet to move the blue beam right or left. Finally, there is the purity magnet and the deflection yoke.

Some screen phosphors are deposited by dusting (Sylvania) where dry phosphor powders are deposited by air on a thin film of photolithographic material. RCA uses the slurry process with phosphor powers premixed in photolithographic emulsion, and then poured on the CRT faceplate. When the tube is put together and installed in a receiver, the three electron beams pass through some 1.146 million holes in a temperature compensated shadow mask to produce either a monochrome or color picture.

RCA's 110-Degree Tubes

Most recent of this manufacturer's offerings are 18, 19, and 25-inch 110-degree color cathode ray tubes that were first seen in late 1970 and during 1971. Fig. 2-10 is a comparison of the standard 90-degree tube outlined in white, with the same 110-degree tube drawn inside in black. Difference in length is 4 inches (17.9 inches compared with 13.9 inches), a substantial saving in receiver depth.

The greater deflection angle requires an increase in the horizontal deflection power needed from 43 percent to 50 percent, and the needed vertical deflection power from 53 percent to 65 percent. The 110-degree deflection angle also introduced additional edge convergence problems. By designing a smaller 29 millimeter neck, however, both power and convergence difficulties came nearer solution, and a more accurate lens and tighter dot-phosphor control brought beam landings on the RGB dot trios under control. Actually, because of the reduction in neck diameter from 36 to 29 millimeters, the yoke measurements were lessened so a 30 percent saving in power was effected, and the short electron gun and closer gun spacing makes all beams pass through a more uniform magnetic field that produces much tighter yoke and tube convergence variations without the addition of further expensive corner convergence correction. The difference between the 110-degree and 90-degree deflection yokes are shown in Fig. 2-11, and both are the familiar saddle type that has been used for some time in the standard 90-degree deflection color tubes.

The 25-inch large screen 110-degree tubes, however, have had one or two different problems. The beam landing and convergence problems that had been readily solved in the smaller size 110-degree tubes were much more critical in the large tubes and required additional design. RCA overcame these problems by inventing what it calls the Precision Static Toroid (PST) deflection yoke shown in comparison with a

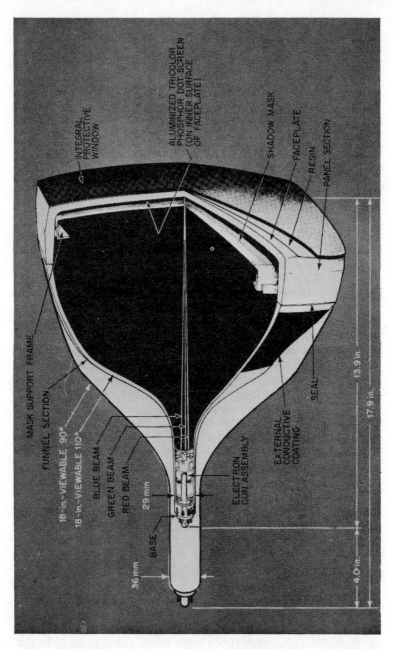

INTEGRAL
PROTECTIVE
WINDOW

ALUMINIZED TRICOLOR
PHOSPHOR DOT SCREEN
(ON INNER SURFACE
OF FACEPLATE)

SHADOW MASK

FACEPLATE

RESIN

PANEL SECTION

MASK SUPPORT FRAME

FUNNEL SECTION

18-in.–VIEWABLE 90°

18-in.–VIEWABLE 110°

BLUE BEAM

GREEN BEAM

RED BEAM

29mm

BASE

36mm

ELECTRON
GUN ASSEMBLY

EXTERNAL
CONDUCTIVE
COATING

SEAL

13.9 in.

17.9 in.

4.0 in.

Fig. 2-10. Comparison of 90- and 110-degree cathode ray tubes exhibiting slim profile of the latter and the thinner, shorter tube neck. (Courtesy C.A. Meyer, RCA)

wide-neck dynamic saddle yoke in Fig. 2-12. As RCA explains it, the PST yoke windings are on a closed ferrite ring core and consist of two coils each that are wound rather fully along the yoke without empty end spaces. By eliminating end turns and using more core, the PST yoke deflects a field that is longer and so bends the three beams more uniformly and gradually with better sweep control. Not only is the PST yoke actually 30 mm shorter, but it weighs 45 percent less than a saddle yoke with the same minimum inner diameter, and also has low-impedance characteristics that are entirely compatible with solid-state receivers. As a result, the spot size and resolution

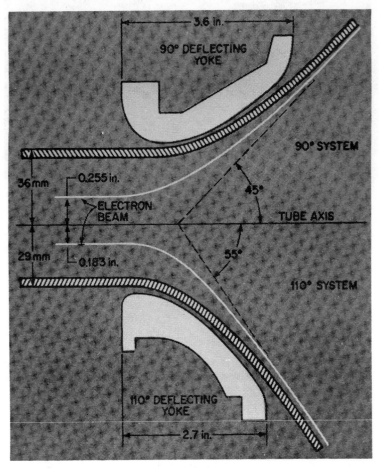

Fig. 2-11. Comparison of the 90- and 110-degree deflection yokes.

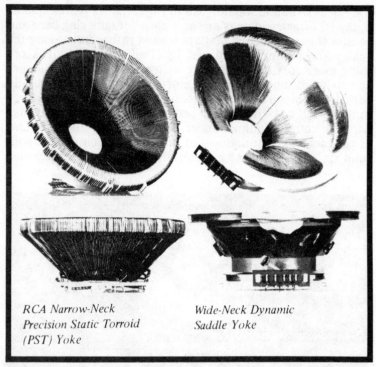

RCA Narrow-Neck
Precision Static Torroid
(PST) Yoke

Wide-Neck Dynamic
Saddle Yoke

Fig. 2-12. RCA's new PST torrodial yoke compared with the wide-neck saddle type. As yoke and tube matches become closer, convergence problems diminish.

of the 110-degree 25-inch tubes that have already been introduced to Europe are considerably better than the 90-degree tubes already on the U.S. market, according to RCA, and the glass transmission, phosphor efficiency and light output are equally proportional.

GE's In-Line Gun

In its smaller screen receivers, General Electric uses a cathode ray tube with in-line RGB guns instead of the conventional 120-degree separated type used in most other U.S. receivers (Fig. 2-13). According to G.E., dynamic convergence correction is already built in because of the way the deflection yoke is wound. The blue and red guns are on the outside, while the green gun is in the center. Normally, in the usual delta tube, all three guns are converged toward the axis or center of the tube. But in the G.E. tube, only the outer beams require convergence.

In Fig. 2-14 you will notice red and blue vertical and horizontal **static** convergence magnets and a pair of red and blue horizontal sliders. There are no green controls, since green is already fixed on the center axis of the tube. In static convergence, one magnet (for each) moves blue or red in a vertical direction, and the other magnet (pair) moves the blue or red beams horizontally. Dynamic convergence is adjusted with red-blue sliders that are connected in series with their respective horizontal and vertical yoke windings, and there is no interaction between either dynamic and static controls, or controls for either outside gun. You will also notice that the purity magnet is positioned between the convergence assembly and the deflection yoke due to construction of the electron guns. Yet purity adjustments are the same except they are made on a green field, probably because green is this tube's reference rather than red or blue. Gray-scale tracking is the same, too, and is initially set with a low brightness level.

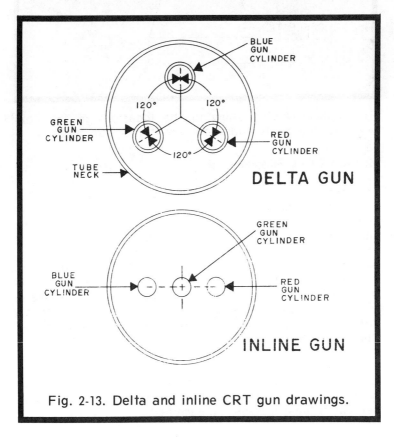

Fig. 2-13. Delta and inline CRT gun drawings.

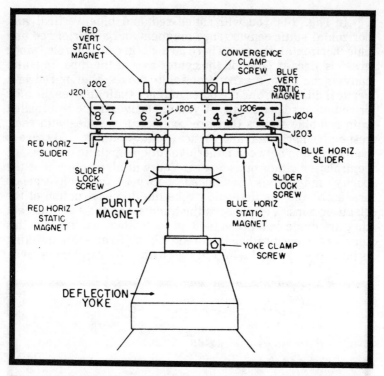

Fig. 2-14. Drawing showing the locations of the convergence magnets and slides for in-line gun tubes.

Zenith's Black Surround CRT

Zenith-Rauland's Chromacolor black surround tube has 1,350,000 phosphor dots deposited at about the same size as the conventional matrix but completely encircled by a black surround material that is highly light absorbent. The iris mask, designed to accompany the dark background process, has apertures that are 9-thousandths of an inch in diameter when the black surround is applied, and are then opened to 16-thousandths of an inch. 450,000 aperatures permit passage of the three electron beams. The diameter of the electron beams is larger than the individual phosphors. So, says Zenith, all "phosphors on the screen can be totally utilized and phosphor efficiencies balanced." In previous conventional 23-inch color tubes, Rauland claims that while the phosphor dots are 17-thousandths of an inch in diameter at center, the electron beams were only 13-thousandths of an inch, with only 45 to 65 percent of the dot areas being excited.

According to the manufacturer, the Rauland 25-inch tube claims a light output of 70-foot lamberts compared to 50.2 for the matrix tube, with a very favorable contrast ratio of 12.1 at 20 ft. candles ambient. The gun for this tube is shown in Fig. 2-15 with its triple electrodes, internal shields and beam dividers. Tubes are hermetically sealed, and protruding pins are solid metal instead of the once-hollow extensions that had to be periodically resoldered for good connections, especially the heater inputs. The long white piece between two of the three guns (bottom of photo) is a glass support bead. The heaters and cathodes are small metal units close to the base (not visible in the photo). The larger round assembly, at the end of the guns away from the base, is grid No. 4, the convergence cage that is affected by both static and dynamic convergence fields.

MOIRE

We are indebted to G.M. Ridgeway, Manager, Applications Engineering, Zenith's Rauland tube division for much of the following information on a topic that will assume a lesser or greater degree of importance, depending on manufacturers' ingenuity and technicians ability. For, during some special operating conditions, a shadow mask picture tube will produce a pattern of brightness variations on the screen that initially has nothing to do with picture content. Something like sheer window curtains with overlapping folds, moire is a pattern composed of both light and dark wavy lines.

The term, moire, was originally applied to a type of mohair ribbed fabric designed to present an appearance of wavy design. This same wavy design may show on the face of a shadow mask color picture tube, resulting in line-to-line variations as the electron beam passes through the shadow mask openings. With continued improvements in phosphor efficiencies and electron gun optics, smaller beam sizes should be realized and, therefore, form sharper raster lines. Consequently, much greater care in such cases is needed during setup adjustments of the newer television receivers to minimize **moire** effects.

Since the invention of the cathode ray tube, experts in the field of electron optics have strived to improve the spot resolution under actual operating conditions. Considerable progress has been made in the performance of the monochrome tube, and until recently it has been significantly superior to that of the tri-color tube.

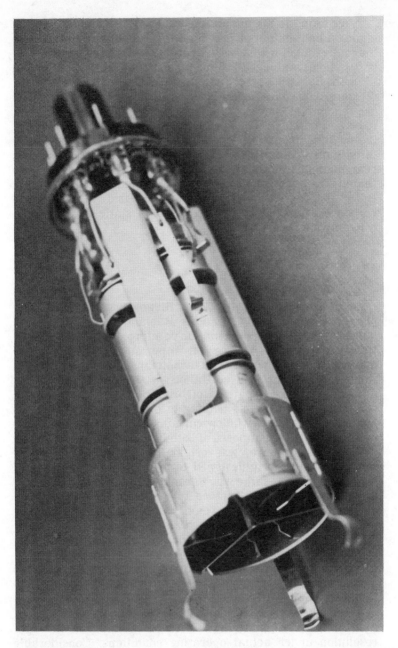

Fig. 2-15. Electron gun assembly for the Chromacolor tube showing the electrodes of the three guns. (Courtesy Rowland.)

Several factors have contributed to the relatively poorer performance of the tri-color CRT gun, some of which are the complexity of manufacture of the 3-gun assembly, the unique delta arrangement which projects three beams—none of which are centered in the deflection field, purity and converged requirements which distort the three beams and the high current requirements for brightness (since over 80 percent of the current strikes the shadow mask, never reaching the screen).

Fig. 2-16 shows typical spot size measurements at various beam currents. The lower curve shows the performance of a typical high resolution gun, while the upper curve shows the performance of an earlier gun. You can see that at any given current, the earlier gun has a much larger spot size.

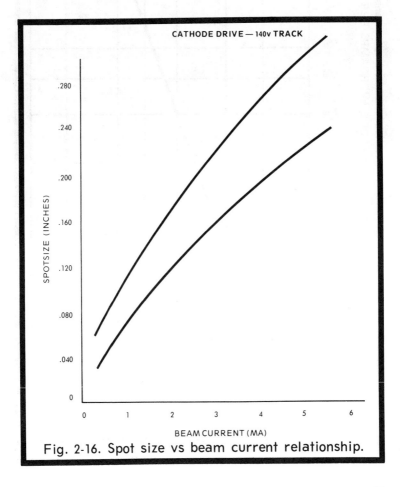

Fig. 2-16. Spot size vs beam current relationship.

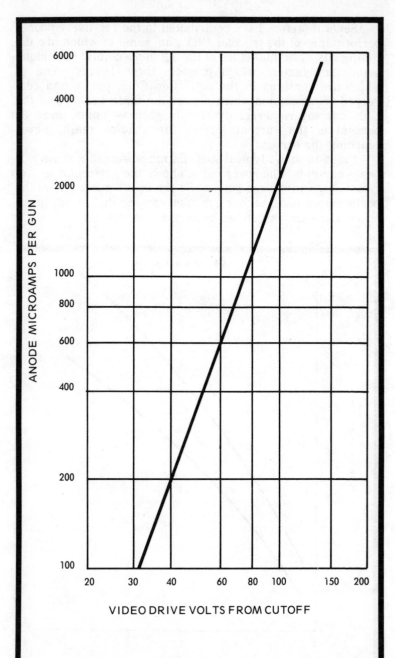

Fig. 2-17. Cathode drive-anode current relationship.

The maximum average current drain from the chassis high-voltage supply seldom exceeds 1500 microamperes which means that each of the three guns is delivering a maximum average current of about 500 microamperes or 0.5 ma. However, the external and internal coatings on the CRT form a capacitor which stores power for higher peak current requirements.

Fig. 2-17 shows the current emitted from each electron gun under various drive conditions. With a setup voltage of 140 volts (G1 to K), the service switch in the setup position and G2 properly adjusted for a white line, you can see that a 140-volt white video pulse will produce about 5.5 ma current from each gun or a total of 16.5 ma. Currents for other video signals can be readily picked off the graph. It is quite normal to have several milliamps current per gun for short durations of time even though the average current is much lower.

Returning to Fig. 2-16, you can now appreciate the concern for the high current spot size. In the 4 to 5 ma range, the upper curve produces a "highlight blooming" condition while the lower curve shows less tendency to bloom. The second area of concern is the 0.25 to 0.5 ma range.

When we focus a tube, we like to see sharp scanning lines. The difference here is even greater. At lower currents, there is about a two-to-one difference in spot size. The improved focus of the tube described by the lower curve is dramatic.

As the electron beam scans across the face of a shadow mask color tube, the beam is striking the metal of the shadow mask a portion of the time and the holes for the remainder. If the diameter of the beam is small with respect to the vertical distance between the aperture holes, it should be possible for the beam to completely miss a hole on successive lines. The missing of holes by the electron beam results in dark areas on the screen which can form into moire patterns.

The modulation of the beam by the holes in the shadow mask will generate a certain amplitude of modulation, depending on whether the beam partially or fully misses a hole. As the beam scans the entire screen, the variations in amplitude will cause a brightness pattern which can be described as a modulating frequency. Depending upon the placement of the scanning lines, it is possible to have a variation in frequency as the beam scans across the face of the tube. These variations in frequency cause the areas of maximum and minimum brightness to form a swirling pattern rather than a straight line pattern. Moire is generally more noticeable without video information, but it should be considered a problem only if it is objectionable on program material at normal viewing distances.

There is a mathematical formula for calculating aperture hole spacing to minimize moire, which we won't go into, except to say that the shadow mask for each color tube family is designed for optimum performance in the 525 scanning line system used in this country. Even so, the optimum design will yield moire if focus quality is sufficiently good.

It should first be pointed out that optimum focus for best picture performance **is not achieved** by focusing for maximum moire! Optimum picture performance can best be achieved by focusing on large area bright spots—peoples' faces for instance. The optimum focus voltage is slightly different at every current level. Focusing for greatest moire will make highlight blooming more objectionable as well as making moire more noticeable. Focusing for best highlight focus will increase low current spot size, reducing moire yet giving the best overall performance. If a picture must be defocused excessively to minimize moire, then the receiver is not properly adjusted or not operating properly! The shadow mask is optimized for moire only if the scanning lines are straight, horizontal, evenly spaced, and properly spaced.

The following approach is recommended to be used on a moire complaint: If it is not objectionable at normal viewing distances, an attempt should be made to explain to the receiver owner that moire is not a defect, but actually an indicator that focus is excellent. Notice that I said, "if moire is not objectionable at normal viewing distances, it is not a defect." Unless the receiver is not operating or adjusted properly, any further attempt to eliminate moire will degrade the picture!

If moire is judged to be excessive, the following parameters should be checked to insure that the receiver is operating as it should:

Anode voltage—Adjust to the design value as indicated in the service manual. Excessive high voltage stiffens the beam and reduces spot size.

Yoke rotation—Throw the service switch and check the tracking setup line. It should be exactly horizontal for minimum moire.

Vertical centering—Also measure the location of the setup line. It should be exactly centered vertically.

Height and linearity—The shadow mask is designed for approximately 10 percent vertical overscan of the raster and equal spacing of all scanning lines. Poor vertical linearity or improperly adjusted height will contribute to moire.

Pincushion adjustment—Using a color-bar generator or other device to give a low-level stationary flat field, adjust the receiver to minimum color level, low contrast, low brightness, then observe the moire pattern. Next, adjust the pincushion control through its full range; stop at the minimum moire point. This point should also be the desired pincushion setting, but should be verified by switching to a cross-hatch pattern.

Interlace—Carefully rotate the vertical hold control while viewing the scanning lines. Pairing of lines will contribute to moire. Good interlace should reduce it.

CRT bias—With the setup switch in the service position, determine the setup bias on the CRT (difference between G1 and cathode voltage). The greater the bias, the higher the screen (G2) voltage will be for a setup line. High screen voltages yield stiffer beams and smaller spot size.

A chassis with a CRT bias control is easily retracked at a lower bias by reducing the bias control setting, then re-setting the screen controls. If the chassis does not have a bias control, a lower bias can usually be obtained on the RGB type chassis by altering a voltage divider to raise the CRT G1 voltage. The color-difference chassis is more difficult to alter. However, by utilizing a signal generator and reducing the brightness to a low level, the three screen controls can be reduced to just cutoff and tracking will be obtained without switching to the setup mode.

Generally, the bias of a large screen color TV set should not be lower than 120 volts. After re-tracking the set, all channels should be viewed and brightness and contrast rotated through their full range to be sure video performance has not been adversely affected (Note: Lowering the bias will increase the available contrast).

The final step is to focus for best highlight performance. Moire should now be eliminated or at least reduced to an acceptable level. It is important to remember that any effort to reduce moire by increasing spot size is accomplished by degrading focus. Care should be taken to insure that spot size is not increased excessively, thereby causing poor focus.

DEFINITIONS

For many, these basic definitions are rather superflous, but they should provide a firmer foundation for those whose knowledge and skills can use a helping hand. Also, for good measure, we'll repeat some of the more uncommon terms at appropriate intervals as we continue through the text.

Luminance Y (brightness) signal: This is the sum of the color signal voltages delivered by the color camera. The composite video (luminance) signal in color transmission is the algebraic addition of red, blue, and green colors sent as a varying voltage proportional to the brightness information from the color cameras. Therefore, the actual signal seems the same as though it came from a single-tube monochrome camera. When monochrome information only is transmitted, the I and Q color signals are absent and, therefore, the receiver color circuits are inoperative, but the luminance (video) amplifiers deliver a black-and-white image of varying intensity to the receiver's cathode ray tube for picture reproduction. Fine detail above 1.5 MHz is transmitted only in black-and-white during any colorcast.

Chroma: Color information produced by the I and Q transmitted signals with respective bandwidths of 0.5 MHz and 1.5 MHz on either side of the 3.58-MHz color carrier. Q colors range from yellowish-green to purple, while I colors contain bluish-greens to orange. The chroma signal also includes the 3.58-MHz color sync pulse of 8 to 11 cycles.

Hue: The chrominance or color element which identifies a scene or object as being red, blue, green and any shades in between. Colors of the same group, regardless of light content, have identical hues.

Saturation: The saturation of any color determines its content of white light. Any completely saturated color has no white light, but unsaturated colors do, and the degree of this saturation is often called color tint. (Your color receiver's tint or hue control is a phase shifter, while the color control regulates the overall intensity (magnitude) of the color signal.)

Color purity: Actually, a field of non-contaminated color. An expression usually denoting clean landing of color beams on the individual color phosphor dots. In many new receivers, the beams now overscan the phosphor dots for maximum efficiency and any excess beam scatter is absorbed by peripheral material called "black surround." If only red dots, for instance, are lighted, then a properly converged receiver should have only a pure red field for satisfactory color purity.

Static convergence: Convergence of the three color beams at the center of the picture tube so that equal registry produces white as the sum of the three red, blue, and green guns. This is done with three small permanent magnets.

Dynamic convergence: The active tracking of the three scanning beams achieved with electromagnets and correction-scanning voltages so that each gun will track linearly from the center of the tube to all four edges and every beam excites only its appropriate red, blue, or green phosphor. Nine potentiometers and three coils control all dynamic convergence adjustments. There is also a lateral static convergence magnet to move the blue beam sideways. And in some of the cheaper receivers, a red-green lateral magnet has been added recently to further simplify convergence.

Color subcarrier oscillator: The color synchronizing signal transmitted on the back porch of the horizontal sync pulse within the 11.1-microsecond horizontal blanking interval. Burst synchronizes the receiver and transmitter during each line blanking interval.

Line scan: The 52.4 microseconds of forward scan time needed to sweep a single line from left to right across the horizontal surface of a cathode ray tube.

Horizontal or line blanking: The 11.1-microsecond interval used for horizontal blanking as the line scan moves back across the tube and into position for resumption of another forward trace.

Field scan: The vertical scan that deflects the swept line downward. There are two fields of 262.5 lines each in every 525-line vertical scan, and each field is completed in one sixtieth of a second, establishing the vertical time of one-sixtieth second or 16.662 milliseconds. Therefore, two fields—a frame—are complete in two sixtieth of a second or one 30th second. So 30 frames per second are transmitted and received. 525 lines x 30 frames equals 15,750 units, and this is the standard horizontal frequency in Hz. Actually, the color horizontal repetition rate is 15,734 Hz, as we've shown already in the color equations.

Aspect ratio: Television receivers have a picture width-to-height ratio of 4:3, and this is termed aspect ratio.

Interlaced scanning: The sequential field blending of the odd and even 262.5 lines in the 525-line scan for a single, relatively linearly picture. The better the interlace, the sharper the picture, since there will be no conflict and no blurring among (or between) lines.

QUESTIONS

1. What are the three characteristics of color and what do they mean?

2. Why is the color TV process called additive?

3. How many colors can the eye recognize, and where in the spectrum do they appear?

4. What combination of colors produce a white raster?

5. Red, blue, and green modulated colors produce what? Saturated and unmodulated?

6. What are the two types of color-bar generators in general use?

7. In gated rainbow and vectorscope color analysis, why is G-Y not so important?

8. G-Y is one of the demodulated chroma colors, so why isn't it used in vectorscope analysis?

9. What single crystal in both transmitter and receiver must have a very close tolerance? What is this tolerance?

10. What is the most closely regulated circuit (component) in both a broadcast station and TV receiver, and how close must this regulation be?

11. Define "purity" simply.

12. What's different about RCA, Zenith, and G.E.'s cathode ray tubes?

13. Define Moire. Do you think it will be more or less prevalent in the years to come?

14. What is the most important point about G.E.'s new in-line tube?

Chapter 3

Receiver Adjustments

Even though manufacturers have improved horizontal linearity immensely over early receivers and have provided height and vertical linearity controls that are no longer interactive, we still encounter sweep circuit failures due to component breakdowns, new deflection yokes, flyback transformers and other high-voltage coils and resistors, and vertical problems of one description or another—not to mention the aging of tube, semiconductor, and passive parts. It's true that a cross-hatch generator can expose nonlinearity equally well or better than a test pattern **over the range you can see.** But you'll find it hard to estimate overscan with a generator, establish the presence of overshoot and ringing, detect alignment smear, or see signal delays called ghosts, because a pattern generator supplies pulses and dot-bars instead of video. Also, when EIA color bars are broadcast, the saturated tones must be processed at specific IRE (Institute of Radio Engineers) levels between white and black references for the receiver chroma circuits to produce excellent color (IRE Standard 50, 23.S1). Then there are VIR and the vertical interval test signal (VIT), telecast by a number of color transmitters throughout the country that, along with an EIA color pattern, can characterize most color receivers just about as fast as you can see them. So for all careful adjustments, receiver analytical analysis, and chroma processing, test patterns and VIT signals are all important.

THE RETMA (EIA) RESOLUTION CHART

Fig. 3-1 is the Electronic Industries Association's test pattern with gray-scale overlays. This chart was originally designed to test television cameras for streaking, ringing, interlace, linearity, shading, aspect ratio, and gray-scale adjustment. But it's equally appropriate for television receivers, except for aspect ratio. In this chart, resolution measurements have been increased from 600 to 800 lines over the original version, and the gray-scale reflection charac-

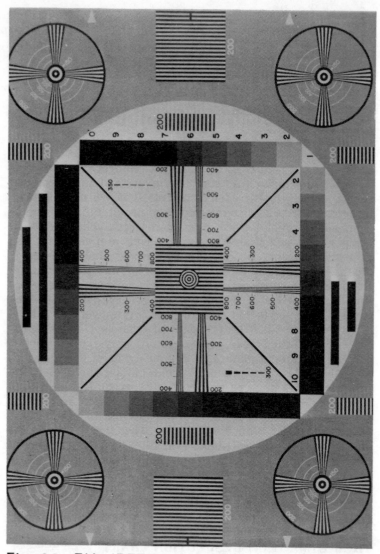

Fig. 3-1. EIA (RETMA) resolution chart for studio camera and receiver setup, full gray-scale shadings, and vertical wedges indicating a 10-MHz bandpass (Courtesy Ed Tingley, EIA).

teristics are improved. In broadcast transmissions, you may or may not see this identical pattern on the air, but one similar enough to it should be available over at least UHF stations during working hours so that its display can be useful. Now, let's learn what it's good for.

The four pairs of vertical and horizontal wedges consist of four tapered black lines equally separated by three white lines. Numbers located at intervals next to the wedge-shaped pairs are the number of lines that can be resolved vertically and horizontally. If you wish to use a simple equation to determine the response frequency from the line numbers beside each vertical wedge, F(frequency) equals N(line number) divided by K(a constant equalling 80). So a 280-line wedge position, then, would amount to:

$$f = \frac{280}{80} = 3.5 \text{ MHz}$$

The response is based on a 63.5-microsecond horizontal trace-retrace interval (one black and one white line) or 250 nanoseconds. And since F(frequency) equals 1 divided by the time base:

$$f = \frac{1}{250} \times 10^{-9} = 4 \text{ MHz}$$

the fundamental video frequency. If the line number on the chart was actually 320, then 320 divided by 4 MHz would be N divided by F equals K, and K would then amount to a constant of 80. So this is what we mean by 4 MHz being the fundamental video frequency. Had you known that line 280 denoted 3.5 MHz, 280 divided by 3.5 MHz would have also equalled the constant K of 80. An EIA list of line numbers and the corresponding frequencies appear in Table 3-1, so the next time you see numbers about the horizontal and vertical wedges, you'll be able to read them:

Vertical Wedge Line Numbers	Fundamental Video Freq.
200	2.5 MHz
240	3.0 MHz
280	3.5 MHz
320	4.0 MHz
350	4.37 MHz
400	5.0 MHz
450	5.62 MHz
480	6.0 MHz
500	6.25 MHz
560	7.0 MHz
640	8.0 MHz
720	9.0 MHz
800	10.0 MHz

Table 3-1. EIA test pattern video response chart.

We hope you remember that these are vertical wedges rather than horizontal wedges, for which the response has been calculated. The reason for this is that **vertical wedges determine horizontal resolution**, and, therefore, the high-frequency response of the receiver. The horizontal wedges, of course, show the low-frequency response and are calculated at a 60-Hz or 16.67-millisecond rate rather than the 15,750-Hz or 63.5-microsecond time of the vertical wedges. And since we're not really interested in the low-frequency response of the average color or monochrome receiver (it should approach DC somewhere), the calculations will be left to you, if you wish to make them. Beware, however, since vertical resolution (horizontal wedges) is very dependent on the size of the scanning spot on the CRT. In addition, 7 percent of the theoretical 525 scanning lines are taken up in blanking, so the perfect receiver would resolve only 490 lines, and most do considerably worse than that. Therefore, it is customary to state only that a receiver will vertically resolve a certain number of lines.

Horizontal Resolution (Vertical Wedges)

Although this is a basic means of determining the upper frequency response of any television receiver, you'll find later that the vertical interval test signal (VIT) is more exact, while a sweep generator modulated by a 45-MHz sine-wave is more precise in bandpass and rolloff evaluation. Should the test pattern fail to supply numbers along broad vertical wedges, select a point midway between its ending and top, call this about 200 lines, and proceed logarithmically from there to 320 lines; a very crude measurement indeed, but something near the ballpark.

Calibrated or uncalibrated, EIA or Indian Head, these test patterns have enormous utility. Ringing at selected frequencies can be seen on patterns with broad vertical wedges. The indication trails the wedge at some line point equivalent to the spurious frequency. In the EIA pattern, the single line of solid, diminishing rectangles in the lower left and upper right portions of the pattern (within the gray-scale bounds) will immediately show ringing, since the vertical wedges cover multiple scanning lines.

Picture IF amplifier regeneration can be detected by fine dark lines crossing the vertical wedges at the frequency of oscillation. Such conditions are especially evident on weak signals where the amplifier gain is high. The frequency of some types of interference might be determined, too.

Therefore, everything from interference, to bandpass, to receiver or even broadcast station deficiencies can be observed and often diagnosed, not forgetting for a moment that vertical size and linearity checks must be made for proper deflection so that the picture aspect ratio remains in its correct proportion (4:3).

Faults Apparent In The Overall EIA Pattern

Shading may easily be checked for a uniform gray background and the 10 steps of gray scale by viewing all four overlays forming a square around the center of the pattern.

Streaking of the top or bottom horizontal black bars indicates low-frequency phase shift or poor DC restoration. These same black bars help adjust camera high peaking circuits to compensate for high-frequency rolloff.

Interlace may be evaluated by single, clean diagonal lines inside the gray-scale square extending outward from the 400-800 line markers.

Improper fine tuning can produce a smeary pattern with poor video definition and often bad sound and sync. A misaligned automatic fine tuning circuit (AFT) will also cause pattern problems.

Corner linearity, for a uniform raster and, perhaps, pincushion settings, can easily be evaluated from the four outer circles, which also indicate corner line resolution. Not much over 280 center lines and probably no more than 200 lines corner resolution is the best to expect from very good receivers.

Secondary images that are reflected and aptly named "ghosts" are very evident in any test pattern. In Chapter 13 you'll learn how to roughly find the origin of these reflections by calculating certain distances.

Determining Interference Frequencies

Ringing, of course, can be easily detected in the EIA photo from the rectangular dashes in the lower upper right and lower left corners of the test pattern, as indicated earlier. But you can determine the exact frequency (f) if you have a good, wideband, triggered oscilloscope. Simply count the number of oscillations, establish your time in microseconds, and use the following equation:

$$f = N \times 10^6/T \text{ (Microsec)}$$

where N is the number of oscillations.

Fig. 3-2. Ten horizontal black lines resulting from overriding the input RF amplifier. Frequency here is 700 Hz.

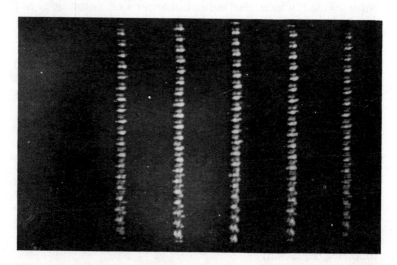

Fig. 3-3. When interference bars are vertical, divide by the line scanning rate of 52.4 microseconds to find the interfering frequency.

TVI: In instances where fine calculations may be difficult, or you simply want to confirm mathematical estimates, TVI (television interference) may be found precisely by using a sweep generator with a 10-MHz sweep width. Uncouple but lay the antenna leads close to the antenna terminals, attach the sweep generator (with internal marker) to the receiver antenna terminals, and take the output off the video detector load resistor. Three pips—the video carrier, generator marker, and interfering signal) will all appear on the RF-IF response curve. Simply tune your generated marker to coincide with the interference and you have an exact frequency measurement. How to get rid of this annoying problem is analyzed in Chapter 13.

But if you wish to find a gross difference interference, but not a fundamental, divide the number of black or white bars by 15.2 milliseconds (approximately the forward vertical trace time) if they're horizontal. In the next two instances, we overrode the input RF amplifier and **did** find the approximate fundamental, but this won't happen often unless your disturbance is overwhelming. With moderate signal strength interference, such signals may be found by matching the number of lines appearing on the screen with those from a calibrated function generator. We'll demonstrate this later. Meanwhile, in Fig. 3-2 there are 10 horizontal black and white bars with a signal generator input of 2 volts, so the frequency (f) would be:

$$f = \frac{10}{15} \times 10^{-3} = 667 \text{ Hz}$$

The **actual** frequency as measured by our Exact 124 multigenerator was 700 Hz, and that's pretty close for such count-and-divide work. If you took the very easy approach and multiplied 10 x 60 Hz, the vertical repetition rate, the answer would have been 600 Hz, a figure that is not as close. So the vertical trace rate in milliseconds is the more accurate, since the full 16.67 millisecond-cyclic rate includes about 1.4 milliseconds retrace.

The same principle is used for higher frequency interference, but with two notable exceptions: The line scan rate is now 52.4 microseconds, and the interference beat or bars are now either diagonal or vertical. To illustrate (Fig. 3-3), we'll put in an Exact sine wave of 1 MHz and see how this figures. Since there are 5 bars:

$$f = \frac{5}{52.4} \times 10^{-6} = 955 \text{ kHz}$$

fairly close to the 1-MHz calibrated signal input. If we had simply multiplied by the horizontal frequency rate of 15,750 Hz, the answer would have been 788 kHz, more than 200 kHz off frequency.

If the number of lines are more than you can count, say in the 50- to 200-MHz region, simply count an inch's worth, multiply these lines by the width of the picture tube and divide by 52.4 microseconds. You should arrive at a relatively reasonable answer, if you have a brute force millivolt or volts input signals.

NTSC COLOR PATTERN

The NTSC color-bar pattern (Fig. 3-4) means a great deal more than simply a few colors following a block of indeterminate white. This or a similar type color pattern is broadcast early every morning by stations all over the country so that transmitter color registration can be aligned to ac-

Fig. 3-4. The NTSC color bar pattern.

commodate the three primary red, blue, and green picture tubes in the camera, and to establish basic transmitter levels prior to transmission. Signals are electronically generated by a color generator, beginning with a gray bar and proceeding through 6 other bars of yellow, cyan, green, magenta, red, and blue—all saturated EIA colors. Color generation may then be monitored by a highly accurate studio-type vectorscope that can measure phase and chroma amplitude to help prevent errors in color transmission.

VECTOR COORDINATES

The vector presentation graphically displays the relative phase and amplitude of the chrominance signal on polar coordinates. To identify these coordinates, the graticule on a Tektronix Type 250 vectorscope (see Fig. 3-5) has points which correspond to the proper phase and amplitude of the primary and complementary colors: R (Red), B (Blue), G (Green); Cy (Cyan), YL (Yellow) and MG (Magenta). Any errors in the color encoding, video tape recording or transmission processes which change these phase and-or amplitude relationships cause color errors in the television receiver picture. The polar coordinate type of display such as that obtained on the Type 520 CRT has proved to be the best method for portraying these errors.

The polar display permits measurement of hue in terms of the relative phase of the chrominance signal with respect to the color burst. Saturation is expressed in terms of the displacement from center (radial length) toward the color point which corresponds to 75 percent (or 100 percent) saturation of the particular color being measured.

The outer boxes around the color points correspond to phase and amplitude error limits per FCC requirements (plus or minus 10 degrees, plus or minus 20 percent). The inner boxes indicate plus or minus 2.5 degrees and 2.5 IRE units and correspond to phase and amplitude error limits per EIA specification RS-189, amended for 7.5 percent setup.

An internally generated test circle matched with the vector graticule verifies quadrature accuracy, horizontal-to-vertical gain balance and gain calibration for chrominance signal amplitude measurements. Two methods of measuring phase-shift are provided. Large phase-shifts can be accurately read from the parallax-free vector graticule. A precision calibrated phase shifter with a range of 30 degrees, spread over 30 inches of dial length, is provided for measuring small phase-shifts.

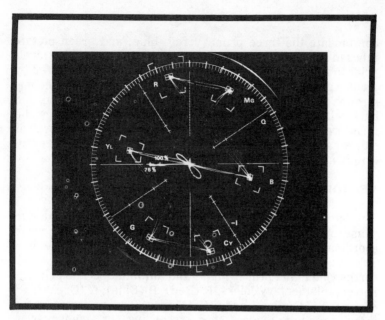

Fig. 3-5. Vector display, full field color bars, 75 percent amplitude, 100 percent white reference, 10 percent setup. Conforms to EIA standard RS189. Type 140 NTSC test signal generator used as signal source. (Courtesy Tektronix)

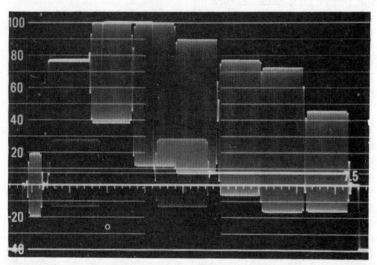

Fig. 3-6. Line rate display of EIA color bars. Notice the levels of the blanking and color burst. Both should always be seen in this display. (Courtesy Tektronix.)

An actual photograph of the NTSC color pattern as seen by a color monitor is shown in Fig. 3-6. Scale at the left is the established IRE scale equalling 7.14 millivolts per IRE unit. And since there are various color levels from -20 to 100, the entire 120 units should be seen in $7.14 \times 10^3 \times 1.2 \times 10^2$, or 0.856 volts or more on the screen of any well calibrated oscilloscope. Observe that yellow and cyan begin at 100 IRE units and extend down to 38 and 12, respectively. Green is 89 to 7; magenta, 77 to -5; and red and blue begin at 72 and 46 each, but stop at IRE -16 together. Burst reference must swing +20 and -20, or 40 IRE units total.

Obviously, transmitters can be adjusted to these values, but can home receivers reproduce these same levels without drastic modification because of distance from the place of origin, terrain, weather, man-made deflection objects, various antennas, good-to-poor lead-in, couplers, and the basic characteristics of the receivers themselves. In making such a picture, we didn't attempt to put an IRE scale on the oscilloscope, but from its calibrated grid markings you should be able to gain a pretty fair idea of what a good home color receiver can do. Remember, however, this picture was taken just after the video detector, and does not represent the color picture on the kinescope. The results, nonetheless, are an exceptional indication of what a home receiver will divulge under these circumstances.

Fig. 3-7 is the photograph of an EIA-NTSC pattern taken at 5:30 AM EDT from one of the broadcast channels in Washington D.C. Transmission distance is almost 40 miles, and there is a 2-way VHF coupler between the receiver transmission line input and the antenna. Receiver is an RCA CTC46. The vertical attenuation setting is 2 volts per division. Compare the perfect pattern in Fig. 3-6A to Fig. 3-7. This is a very tough comparison because of transmission, reception, etc., but notice how closely it compares with the original picture? The almost identical levels, the mostly very clean cutoff of the various color frequency oscillations, the I and Q positions, color burst, relative amplitudes, and all characteristics are there to be carefully observed. What you will really find is that with this color pattern and the VIT test signal, you may characterize almost any commercial television receiver in very short order, and do it technically and not subjectively.

In the future, should you use this pattern to any extent (depending on the time it's broadcast), you will find that it will tell you immediately if the receiver needs alignment, how the

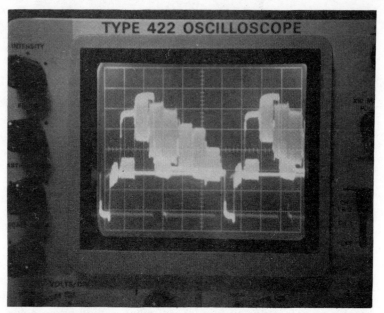

Fig. 3-7. Photo of the EIA color pattern as seen on an RCA CTC46.

overall color is processed (at least through the video detector), how good the burst amplitude is, the Q and I placement portions, and the shape of the horizontal blanking pulse. Be sure you check fine tuning, however, since this can always have a radical effect on positioning of the 50-percent chroma-video IF response points.

VERTICAL INTERVAL TEST SIGNAL (VIT)

The Federal Communications Commission permits transmission of test signals between lines 17 through 20 of the vertical interval of each field for such checks as reference modulation, system performance, cue and control signals. Most all broadcasters use VIT signals with variations, but all usually include multibursts, white-and-black reference lines, a large rectangular window pulse, a sine-squared pulse, and either a modulated or unmodulated staircase. (As of April 1, 1972, the VIT signal was **required** for remote transmitters.) You may quickly judge relative amplitudes, white and black levels, the horizontal blanking interval, burst, gray-scale and differential phase. The window signal is used to gauge transient response, distortion, ringing, etc. If drawn in strict line sequence, the signal would look like it does (if perfect) in Fig.

3-8. Add a modulated staircase of the color bar between lines 15 through 21 and the burst phase, together with all other subcarrier frequency components of the test signal outputs, may be checked for phase variation. If a modulated staircase signal is inserted on lines 18 or 19, or a 75-percent amplitude, full-field color-bar signal included, the entire video signal can be tested while programs are being broadcast. The VIT signal, then, is thoroughly useful in the **studio**, obviously. But what about in the receiver?

The vertical interval test signal arriving at the receiver is probably not particularly affected if the antenna transmission lines, and 2- or 4-set couplers are of good quality. At least that's the preliminary indication we have. So you actually have a very solid view of what an individual receiver is capable of, simply by looking carefully at the VIT on the graticule of your oscilloscope.

The photograph in Fig. 3-9, was taken at 2 volts vertical per division and a sweep of 10 microseconds per division. The scope can't possibly separate lines well in such a sequence, so the following staircase gray level is superimposed over the rectangular window. The only actual peculiarity we notice is that there is a 0.75-volt spike at the end of what would be the 4-MHz multiburst oscillation, and once more just before the window. This is probably some induced transient used for special peaking, appearing at 30-microsecond intervals. On

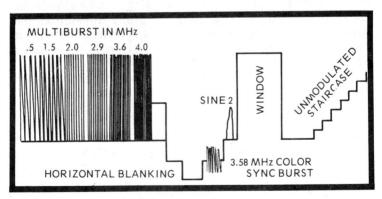

Fig. 3-8. Vertical interval test signal (VIT) transmitted between lines 17 and 20 of the vertical scan interval; another aid in checking receivers for proper operation, alignment, etc. Good resolution of this pattern takes a better than average scope—no AC recurrent sweep types. It must resolve horizontal lines cleanly at 10 microseconds per division.

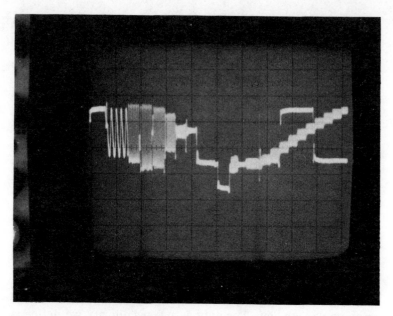

Fig. 3-9. Actual photo of a good resolution VIT on an RCA CTC46. Bandpass is 3.5 MHz.

another channel (Fig. 3-10), the spike was less pronounced and the multiburst amplitudes much more even in amplitude, but the window signal was not apparent in the transmitted pattern. So there are undoubtedly both transmission and reception variations that need to be taken into consideration, but there shouldn't be any difference in the fundamentals. If there is any real doubt, simply compare one receiver with another, and the comparison will then become relative.

If we were to analyze Fig. 3-9, we'd find that the RCA CTC46 shows a good bandpass down to 3.6 MHz with a little rolloff, possibly some high-frequency response peaking, no amplifier ringing, good differential gain characteristics with a fairly linear modulated staircase. On the second channel checked, the multiburst did **not** exceed the amplitude of the white reference, and the five discernable frequency bursts were somewhat more uniform in height. Obviously, the receiver is not in need of alignment. Its quality is much better than average, and with such a pattern you can be certain of excellent operation. Staircase nonlinearity, we're told, is often due to network transmission and modulation by the 3.58-MHz signal. An unmodulated staircase should be straighter and considerably more linear than the two shown.

In checking with broadcasting stations, we learn that VIT patterns are somewhat variable, in that the network may run some **amplitudes** in the multiburst higher than others at various times, and that some standard signals may or may not be included during the various telecasts. Therefore, if you are concerned over the appearance of any VIT signal, simply phone the station telecasting the channel to which you're tuned and ask engineering to look at the VIT signal coming off the network and the one that's locally transmitted. We found they were happy to comply.

Another caution in using these patterns: You may find that several well-known sets don't show more than a 2.8-MHz IF passband. This is said to be due to any number of subjective considerations, one of which may be a "smooth" picture. Nonetheless, this does cut down on fine detail in the picture. A 2.8-MHz IF receiver obviously won't pass the band of frequencies that a 3.6-MHz IF set will. In addition, always check the fine tuning before studying the VIT or color pattern, and be sure your antenna is not loaded with couplers, SWR, and a poor quality feeder line.

As a further aid to understanding the VIT and other signals, even though these are used almost entirely for studio

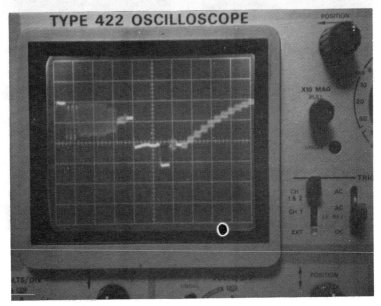

Fig. 3-10. Second station VIT; not quite as outstanding, and with no window.

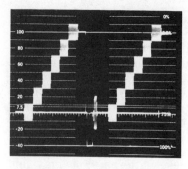

Modulated staircase signal. 2 H SWEEP, FLAT response.

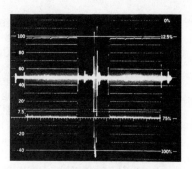

Modulated staircase signal. 2 H SWEEP, DIFF GAIN response.

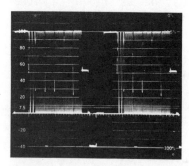

Multiburst signal. 2-H SWEEP, FLAT response.

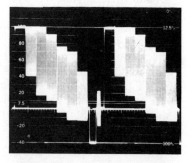

75% saturated color bar signal. 2-H SWEEP, FLAT response.

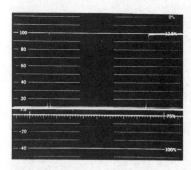

.125 μs HAD Sin² Pulse and Bar. 1-μs/div calibrated sweep, FLAT response.

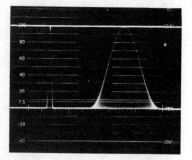

.250 μs HAD Sin² Pulse (left) and Modulated 20 T Sin² Pulse (right) with superimposed Bar Signal (top).

Fig. 3-11. Various NTSC signal generator displays. (Courtesy Tektronix)

system checks, Fig. 3-11 is a series of patterns generated by Tektronix NTSC signal generators in various generating modes.

THE VERTICAL INTERVAL COLOR REFERENCE (VIR)

It would be a definite advantage if black reference, sync, color burst, and especially the phase and amplitude of the chroma signal received automatic compensation at the broadcasting studios **before** transmission. No more hue and chroma amplitude adjustments, not even when a switch is made from camera to camera or when a commercial comes along with the expected magentas and cyans creeping in everywhere. The broadcast companies and the FCC Field Tests Subcommittee, headed by Eric M. Leyton, are all working on the **manually** controlled initial stages of this development now (1972).

The VIR signal (Fig. 3-12) is expected to be added to the broadcast signal at the output of the studio control room and placed on line 20. Substantially, as in the beginning, and as shown in the illustration, the entire sequence will be contained in one 63.5-microsecond forward scan and retrace line. Its order and content are as follows: The interval from the beginning of horizontal sync pulse to the end of blanking level is 12 microseconds, where the color burst is also included. The signal then rises to a level of 50 IRE units where the chroma reference bar is placed between 50 and 90 units, with 70 IRE units as the center reference. This bar lasts for 24 microseconds, and is then followed by the luminance reference at 50 IRE units for 12 microseconds. Following this is a sharp downward transition to the black reference level at 7.5 IRE units that also lasts for 12 microseconds. Afterwards, the blanking level begins, followed by the sync pulse and another burst.

Should this or a similar station-transmitted signal be adopted as a national standard, broadcast stations would be required to keep their transmitters adjusted to such references at all times, and receiver manufacturers could then design somewhat more closely so that many of the color and luminance problems we have today could be largely eliminated.

Further, should the VIR signal—like the VIT signal—be **required** as the VIT currently is for broadcast remotes, field personnel would have a highly reliable means not only to adjust and repair faulty sets, but could easily evaluate receivers on the basis of these two or three (the EIA color

93

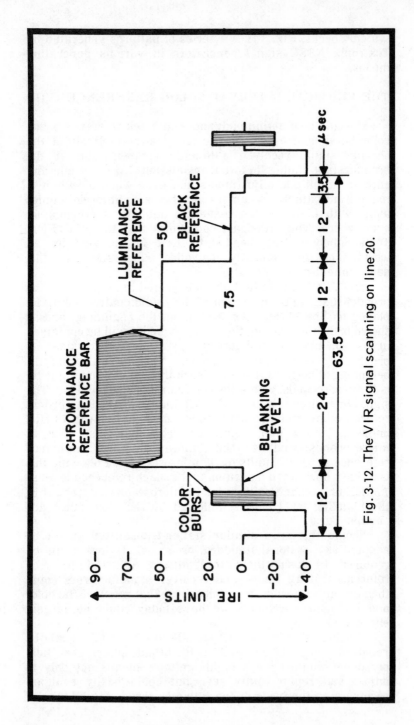

Fig. 3-12. The VIR signal scanning on line 20.

pattern) signals so that future "eyeball" guessing would be eliminated when the new lines appear.

ADJUSTMENTS

Automatic Gain Control (AGC). Some receivers, such as RCA's CTC 46 chassis, do not have an automatic gain adjustment, while others have two. The normal IF AGC setting usually determines the DC bias to the first and second video IF amplifiers, limiting the operation of these stages so that neither they nor the following third video IF or subsequent video amplifiers will overload and cause "sync bends" and picture distortion. Severe cases of AGC maladjustment will even produce negative pictures, where whites and blacks reverse in the scanning scene. The ordinary color receiver has a transistor-resistor or diode-resistor isolation or voltage drop system, called a "delay," so that the RF amplifier will be controlled only when the IF source is fairly full-on. Other receivers have an additional RF control, such as Motorola, that can be set simply by watching the picture for proper signal gain. The IF AGC is always set on the strongest received channel.

Noise Control. Noise controls are found only on the more deluxe receivers and require additional circuitry to function. The purpose of a noise control circuit is to cut off sync and often video for several nanoseconds or perhaps microseconds when a transient of noise from some source enters the video IF amplifiers. Receivers without noise circuits are usually disrupted for the duration of such an undesirable spike of spurious voltage. The noise adjustment often is related to the AGC adjustment, and you may have to set the noise circuit bias first before regulating the AGC.

Fine Tuning. The fine tuning adjustment simply varies the oscillator frequency. But fine tuning of the receiver's local oscillator also moves the video, sound and chroma response around the receiver's IF passband, and can very easily detune the entire picture or cause interference between sound, video IF carrier, or chroma that will result in all sorts of undesirable distortion. Especially is this true of the 920-kHz beat between the 4.5-MHz intercarrier sound and 3.58-MHz chroma subcarrier which produces annoying fine lines in the picture. Without picture distortion, these same fine lines (but at different repetition rates) can also come from improper filtering of the 3.58-MHz receiver-generated subcarrier following the R-Y, B-Y, G-Y demodulators. Of course, this can't be tuned out and is a clue to telling the two disturbances apart. Also, when fine tuning affects **both** definition and probably contrast, the set may need alignment or, at least repair.

Horizontal Linearity, in vacuum tube circuits, usually shapes the current in the damper-flyback transformer primary for the best linear sweep. In at least one model solid-state color receiver, this control is in series between the side pincushion transformer and a primary winding on the flyback transformer. The other end of same flyback winding is connected directly to the receiver's high-voltage adjust. Current through the linearity coil, then, is adjustable to aid the pincushion transformer and deflection yoke in linearly controlling horizontal scan. In the CTC 46, the older vacuum tube Motorola receivers and several others, the linearity control is factory preset and need not be adjusted unless a new HV transformer, pincushion transformer, or deflection yoke is installed. Then, you may or may not have to adjust it (by distributor or manufacturer's instructions), depending on how the resulting pattern appears. As a reminder, in all linearity, height, or width adjustments, use a crosshatch signal first for symmetry, then check the raster size.

High-Voltage Adjust. Often, this is a simple dark current (CRT) setting using a calibrated meter, high-voltage probe, and no raster. In some of the new receivers, the high voltage, like the linearity coils in others, is already factory preset. And in Zenith's new solid-state receiver, there is **no** high-voltage adjustment. In many older receivers, it is necessary to measure the horizontal output tube cathode current while adjusting an efficiency coil so the HV section operates within the prescribed 23 to 27 KV limits. And in those receivers with high-voltage rectifier tubes and shunt regulators there is the possibility of exceeding the 0.5 milliroentgens prescribed as the maximum safe level of x-ray radiation by the National Center for Radiological Health. Newer receivers, especially the solid-state sets, have a different means, including brightness limiters, of handling excess CRT beam current and high voltage. If there is a horizontal output tube cathode current measurement, it is for a minimum, and certainly around 200 milliamperes. The efficiency coil, you may know or remember, is a means of linearizing the deflection yoke current as it passes through zero between the small overlap interval when the damper tube cuts off and the horizontal output turns on. After the efficiency adjustment is completed, of course, the high-voltage potentiometer is set with the HV probe in contact with the anode of the cathode ray tube (brightness extinguished).

Vertical Height & Linearity. It used to be that servicemen were continually cautioned that **any** adjustment of vertical height and linearity would cause convergence and purity shifts. Now, one manufacturer simply cautions that "large excursions may affect dynamic convergence. Adjust before converging set." In other sets, especially one monochrome version, the height and linearity controls are entirely independent of one another and adjustment is no longer interactive as it is on most receivers.

In the tube receivers, the vertical height often adjusts the DC bias on the vertical output tube grid, thus varying the drive to the tube and so causing an increase in the vertical amplitude of the output waveform. The vertical linearity control limits the plate current of the vertical oscillator, and so affects the charging time of the time-constant capacitors in the oscillator circuit and, consequently, the forward trace linearity of the deflection yoke current. Height usually affects the bottom of the picture, while linearity seems to work principally on the top. However, in most receivers they are interactive, since they are part of a single oscillator-output triode-pentode combination tube circuit. In semiconductor sets, the height is found in the collector of the vertical oscillator, and the linearity control often varies the feedback from a push-pull complementary symmetry transformerless output circuit to the base of a pre-driver stage following the vertical oscillator.

Other adjustments such as master brightness, AccuMatic, Tint Guard, Instamatic, AFT, deflection yoke tilt, pincushion correction, screen, CRT bias, background, Color Commander, color temperature, horizontal frequency, etc., are either special settings or color adjustments that are unique to each receiver. The hue (tint) and chroma controls are more or less individual for each receiver, except that the hue control must move the color pattern (as generated by a color-bar generator) at least 30 degrees on either side of the center setting, so that fleshtones are normally seen on broadcasts when the control is in its mechanical center, with greens on the left and purples on the right when turned to either extreme. Automatic tint control and sepia shading are two other settings that are related mostly to one series of receivers.

We also might include a color killer adjustment that some receivers still retain and which is simply tuned for minimum color snow on any off-channel setting. There is, too, a high-low power transformer line voltage setting that adds or removes additional turns on the primary of the power transformer so that the receiver operates as closely as possible to the normal standard of 117-120 volts—its design center. It's wise to check the line voltage with a meter if there's any possible doubt.

THE NBS NATIONAL TIME CHECK

As this book is written, the National Bureau of Standards, Boulder, Colorado is working on an experimental time-code system that could give every television owner in the U.S. a precise time check or clock set within 2 or 3 microseconds over most of every 24-hour day, or for at least the time your local television station remains on the air.

This proposed time- and frequency-check system operates on a 1-MHz carrier during the scan portion of line one and is decoded at the receiver by a phase-locked oscillator, somewhat similar to the regenerated color subcarrier in any color receiver that uses a phase correction loop. This means you would have a 1-MHz test frequency and time signals as well. For TV stations, an entirely independent "alphanumeric message processor and video display unit" is included as part of the proposed system. Here, messages can be transmitted during the line 1 intervals not used for the time and frequency signals, which occupy the first four lines number 1 (there's a number one line each frame) of each second, leaving 26 first lines for other information transmissions. Messages up to 512 characters long may be sent in approximately 15 seconds. The TV station's time would be automatically corrected, plus there would be a one-way message channel of some 400 words per minute, a video title generator, a 1-MHz frequency source and a 1 pps pulse train. The system has already been field tested in Denver, Cheyenne, Los Angeles, and Washington, D.C.

For the standard home receiver, a single integrated circuit will decode and set an electric clock, and with additional circuits it could be made to produce binary coded decimal signals for hours, minutes and seconds that could appear at some point on the CRT face during any period manually selected by the home owner. There is also a means for identifying the individual time zones, with automatic corrections for daylight savings.

NBS warns that existing receivers are not "easily" modified for such displays, and recommends that any such IC or circuit and clock be installed by the manufacturer **before** the new receiver leaves the factory.

QUESTIONS

1. In color TV evaluation, name the two most valuable test patterns.

2. Calculate frequency in vertical test pattern wedges if the constant is 80; the line-wedge response is 195 lines.

3. When you "read" vertical test pattern wedges, are you looking at high- or low-frequency response?

4. Name some of the problems that can be identified by viewing an EIA monochrome pattern?

5. What instrument can exactly determine interference frequencies? How do you do this mathematically?

6. A vectorscope is used basically for what?

7. What is a good way to determine if a set needs alignment without using a sweep generator?

8. Are noise and AGC adjustments ever related. Which is adjusted first?

9. Why is fine tuning so important?

10. Do all receivers have high-voltage adjustments?

11. What adjustments were originally thought to cause large purity and convergence shifts?

12. How do you adjust the color killer?

Chapter 4

Purity, Gray-Scale & Convergence Procedures

The three guns in any color tube of the usual shadow mask variety are set 120 degrees apart from one another. Since they emit separate beams, each must initially be converged at the center of the CRT to form a single, uniform pattern. And if the right percentage of red, green, and blue beams are made to land on the   RGB spots over the entire screen, the pure output with no incoming signal should be a clean, greyish white raster without contamination or shadows. Should one of the beams be tilted away from the other two, distinct fringes of the color produced by the displaced gun will outline larger objects or persons in the picture. Quite simply, this is called color fringing, and is objectionable on both black-and-white and color telecasts because it is distracting to any viewer.

COLOR PURITY

There is also the problem of color purity. If, after a receiver has been either manually and-or automatically degaussed, there is a medium-to-large splotch of undesirable color on the screen, this is an indication that the beams are striking other than their normally assigned phosphor dots, thus creating some undesirable primary hue that is called an impurity. Any adjustment to correct the condition is made only after connecting an RF dot-crosshatch generator (Fig. 4-1) to the RF antenna terminals. Then, the red, blue, and green static magnets (Fig. 4-2) and the blue lateral magnet are positioned for best center convergence as demonstrated by a series of vertical white dots or a bar in the middle of the raster. Then, the signal generator gun killer switches or screen controls are positioned so that two of the three CRT guns are cut off, usually leaving red (the previously least efficient phosphor, before 1969-1970) as the overall color for final adjustment. When this is done, locking hardware for the deflection yoke is loosened (Figs. 4-2 and 4-3) and the yoke is slid back toward the convergence assembly without either disturbing the convergence magnet assembly or tilting the deflection yoke. Tabs on the purity ring are then adjusted so

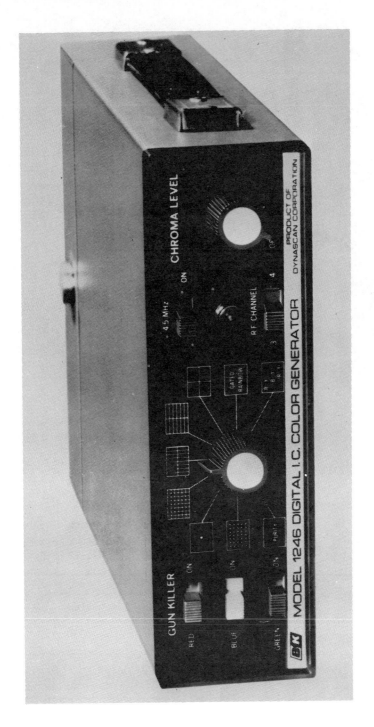

Fig. 4-1. B & K Model 1246 digital count-down IC color-bar generator.

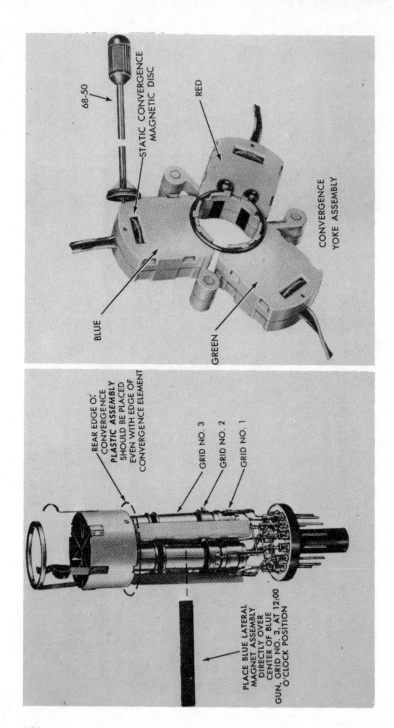

68-50

STATIC CONVERGENCE MAGNETIC DISC

RED

CONVERGENCE YOKE ASSEMBLY

BLUE

GREEN

REAR EDGE OF CONVERGENCE **PLASTIC ASSEMBLY** SHOULD BE PLACED EVEN WITH EDGE OF CONVERGENCE ELEMENT

GRID NO. 3

GRID NO. 2

GRID NO. 1

PLACE BLUE LATERAL MAGNET ASSEMBLY DIRECTLY OVER CENTER OF BLUE GUN, GRID NO. 3, AT 12:00 O'CLOCK POSITION

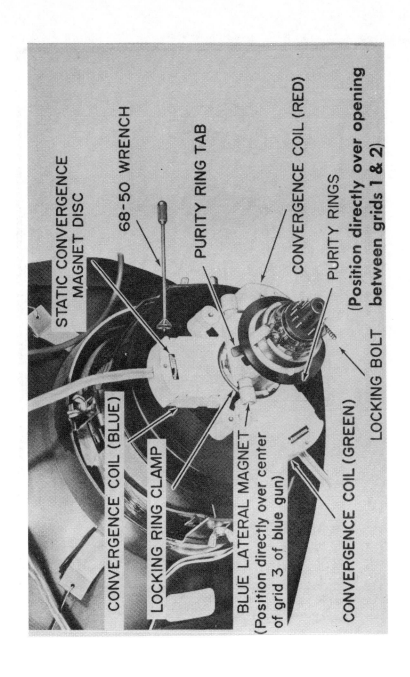

Fig. 4-2. Zenith's CRT gun and static-dynamic yoke assembly for both hybrid and solid-state receivers.

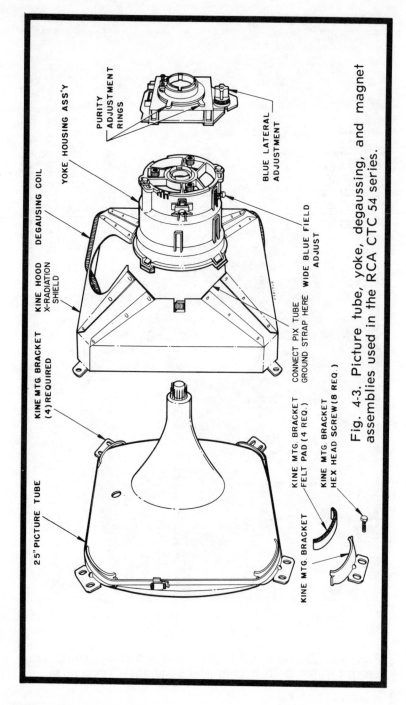

PURITY ADJUSTMENT RINGS

BLUE LATERAL ADJUSTMENT

YOKE HOUSING ASS'Y

DEGAUSING COIL

KINE HOOD X-RADIATION SHIELD

KINE MTG. BRACKET (4) REQUIRED

CONNECT PIX TUBE GROUND STRAP HERE

WIDE BLUE FIELD ADJUST

25" PICTURE TUBE

KINE MTG. BRACKET FELT PAD (4 REQ.)

KINE MTG. BRACKET HEX HEAD SCREW (8 REQ.)

KINE MTG. BRACKET

Fig. 4-3. Picture tube, yoke, degaussing, and magnet assemblies used in the RCA CTC 54 series.

104

that the purity magnet field cuts through the electron beams passing through the neck of the cathode ray tube, moving them at right angles to the field. When the tabs are spread (Fig. 4-4), the magnetic field strength increases and so does the amount of deflection; rotating the entire magnet moves the beams in a circular path. As the tabs are adjusted, a red ball (Fig. 4-5) appears on the center of the receiver. Thereafter, the deflection yoke is slid forward until a uniform red field covers the entire face of the cathode ray tube. This is an actual adjustment of the deflection center near the bell of the tube. By alternately turning on and off the red, blue and green beams, the other two pure fields may be verified also. Observe that the position of the yoke does **not** affect the electron beams that pass through its center and are not deflected, but only those spreading from the larger middle area to the edges of the rectangular CRT.

Should contamination still appear on certain portions of the screen, a manual degaussing should be undertaken to demagnetize the shadow mask and other metal objects in the screen area. If irregularities still persist, then either the purity magnet is defective, the degaussing coil is not doing its job (you might try it on another receiver), or the CRT shadow mask itself and-or phosphors are permanently affected. Occasionally, you may find a color receiver that has been placed close to a large current-consuming appliance such as an electric cook stove, oven, etc. Such appliances will radiate enough AC to magnetize shadow mask and other metal in the screen area. The only solution is a thorough manual degaussing. In the newer receivers, however, automatic degaussing, which occurs when the receiver is turned on should nullify any such occasional problem.

GRAY SCALE

While many prefer to converge a receiver first, then proceed to gray-scale tracking, we think it's sometimes better to adjust the CRT biases immediately after purity checks, since you should have already done a static convergence during purity adjustments, leaving only a dynamic touchup as the final setup step. In this case, it's a matter of simple preference, and the order really makes little difference unless the receiver is so far out of convergence—a real rarity—that overlapping colors will give you problems in obtaining a gray-white screen. Should such a condition arise, there would be no alternative but to undertake convergence first.

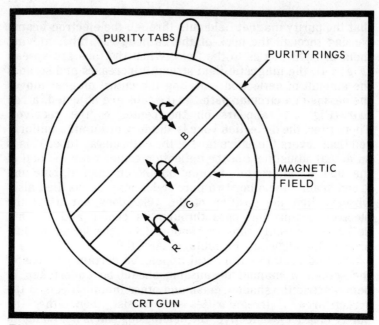

Fig. 4-4. Moving the purity ring tags together or apart decreases or increases the field strength, respectively, while rotating the purity ring moves the electron beams in a circle. Center dots are the beam starting points before adjustment.

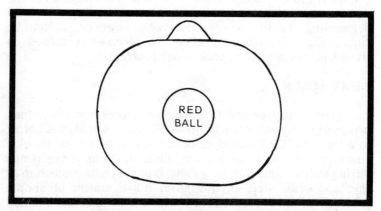

Fig. 4-5. A red ball appears as the tabs are adjusted for center purity, then the yoke is moved forward for a uniform red (or blue, green) field over the entire face of the CRT.

Gray-scale adjustments vary somewhat from receiver to receiver (Fig. 4-6), and the only way to be certain of an absolutely correct procedure is to consult one of the TAB color manuals or the actual manufacturer's literature. What you're trying to do in gray-scale tracking is to set and maintain the three CRT gun beam currents at the same **ratios** throughout both the range of the brightness control **and** any and all incoming luminance signals. Beam currents are to be fixed at certain relative values after the cut-off (extinguish) point for each beam is determined. The actual CRT screen temperature sought is about 9300 degrees Kelvin, where K equals degrees Celsius + 273, and degrees Celsius equals five-ninths (Fahrenheit -32).

Color Temperature Adjustments (RCA)

1. Preset:
Screen controls to minimum (CCW)
Drive controls maximum (CW)
CRT bias control minimum (CCW)
Brightness control maximum (CW)
Contrast control at center of range

2. Set the service switch to **service** and advance the screen controls until each gun barely lights a color line on the CRT. If any gun does not light, rotate the bias clockwise until all three guns show faint red, blue, and green lines. It's preferable to work in a darkened room.

3. Set the service switch to **raster** and adjust the CRT drive controls for a gray-white (9300 degrees K) raster. One drive must be maximum upon completion of setup.

4. Rotate the brightness control over the total range and check the low and high lights. If tracking is poor, dim the raster and readjust the drive controls. The service switch is now set in the **normal** position.

It may also be necessary to touch up the pincushion adjustments for straight horizontal lines at the top and bottom of the raster. This takes in both amplitude and curvature (phase) corrections, and is done with a simple crosshatch pattern from a color-bar generator. While a crosshatch pattern is present, check for vertical and horizontal width and linearity and undertake any adjustments needed, making sure the raster is centered.

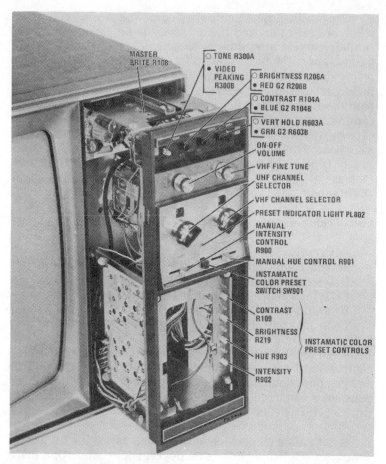

MASTER BRITE R108
TONE R300A
VIDEO PEAKING R300B
BRIGHTNESS R206A
RED G2 R206B
CONTRAST R104A
BLUE G2 R104B
VERT HOLD R603A
GRN G2 R603B
ON-OFF VOLUME
VHF FINE TUNE
UHF CHANNEL SELECTOR
VHF CHANNEL SELECTOR
PRESET INDICATOR LIGHT PL802
MANUAL INTENSITY CONTROL R900
MANUAL HUE CONTROL R901
INSTAMATIC COLOR PRESET SWITCH SW901
CONTRAST R109
BRIGHTNESS R219
HUE R903
INTENSITY R902
INSTAMATIC COLOR PRESET CONTROLS

Fig. 4-6. Motorola's Quasar II front panel control locations, Chassis STS-934.

In actual receiver adjustment, you must find a cut-off point for each gun that will allow all three beams to extinguish at the same single low setting of the brightness control. Then the relative beam currents are controlled by advancing or retarding the various voltages at the grids and-or cathodes of the cathode ray tube, depending on the bias system used. Sometimes these voltage controls are identified as background controls or they may be called gain controls, but both serve the same purpose: that of matching the beam currents to the various phosphors used over the range of both the brightness control and various voltage amplitudes of the received monochrome signal. Careful setup insures that the color signals will then be reproduced within receiver limitations. When the cathode ray tube is perhaps a bit low on output, there is often another potentiometer in the grid circuits of the newer cathode-fed picture tubes that can be adjusted to usually produce a lower cut-off voltage so the grid 2 voltages can be effective over large settings.

During general gray-scale adjustment procedures, the screen voltages are set to virtual cutoff and the control grids (or cathodes) are biased at some relative setting. To make the exercise more realistic, we'll use directions supplied by RCA for the CTC46 and CTC49 receivers (Fig. 4-7) as an example. When these things are done you are ready for the usual combination static-dynamic convergence.

STATIC-DYNAMIC CONVERGENCE

The convergence adjustments are used to **aim** the CRT electron beams so that three pictures are superimposed over one another to produce a single unified image with maximum detail in monochrome transmissions and the necessary color mix in chroma transmissions. Static convergence is achieved by small adjustable permanent magnets in the external portions of the convergence yoke assembly (Figs. 4-2 and 4-8) several inches above each gun. Lines of force resulting from the static magnets reach the individual pole piece located **within** each gun. As each magnet is moved, the beams shift in the direction of the arrows in Fig. 4-8. The red and green beams, therefore, must be statically converged first, then the blue beam moved down to intersect the other two. Should the blue beam require side-by-side adjustment, there is also a blue lateral permanent magnet that moves the blue beam horizontally, left or right, as needed.

Because of the geometry of the picture tube, beams deflected to areas other than the center of the screen are subject to further misconvergence because they have further to go to reach the screen. To correct the situation, varying electromagnetic fields are needed to converge the beams at all points on the screen. So further down in the yoke assembly, close to the neck of the CRT, are three electromagnets placed directly over the three RGB guns. The magnets consist of horseshoe-shaped ferrite cores about which are wound two pairs of coils in series. One pair corrects for misconvergence vertically and the other corrects for misconvergence horizontally. The correction fields are generated by a 60-Hz vertical AC voltage applied to one set of coils and a 15.75-kHz

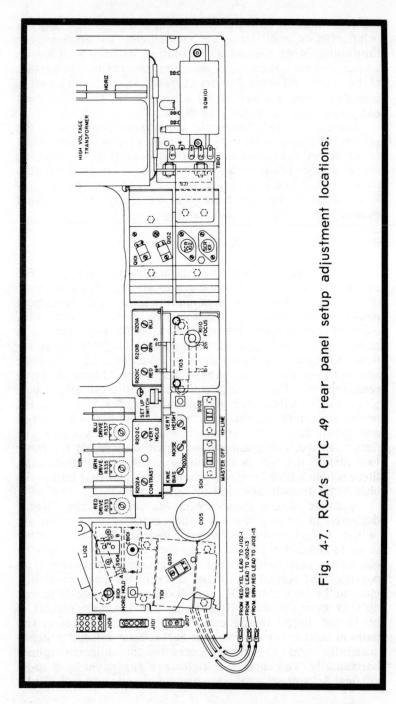

Fig. 4-7. RCA's CTC 49 rear panel setup adjustment locations.

horizontal AC voltage fed to the other set of coils. Since the electron beams of the three guns are sweeping the CRT screen, they must pass through the magnetic fields set up across the pole pieces by each two pairs of coils. The fields change according to the AC applied as each beam is deflected simultaneously at both horizontal and vertical scan rates, and are positioned on the raster at points proportional to the instantaneous strength of the varying magnetic field. Since the strength of the field changes as the three beams move across the face of the CRT, this method of active sweep correction is called dynamic convergence.

In some receivers, more than others, dynamic and then static adjustments must be made, sometimes alternately since there is interaction in most cases, to gain an acceptable overall degree of convergence without excessive fringing. There's probably no such thing as perfect convergence, but an effort should be made to keep misconvergence at a minimum toward the center of the CRT and permit only small discontinuities at the top, bottom, and extreme left and right edges. In this way, fringing can be checked, and most viewers will consider the adjustments satisfactory.

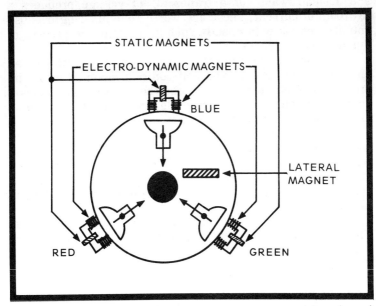

Fig. 4-8. Drawing of the convergence assembly placed over the three pole pieces inside the CRT gun. The arrow shows the direction of beam movement.

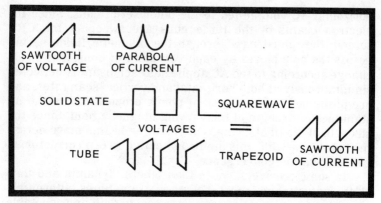

Fig. 4-9. Dynamic convergence currents developed by voltages from the horizontal output circuits.

Dynamic Convergence Controls

In dealing with output and convergence circuits, remember two important rules (Fig. 4-9): a sawtooth of voltage through an inductor produces a parabola of current, and a squarewave (or a trapezoid) of voltage produces a sawtooth of current. These are worth remembering because you are dealing with exactly such currents in the process of dynamic convergence.

Many of the older dynamic convergence assemblies had an even dozen resistor and coil controls mounted on an individual printed circuit board, and there were four separate controls called vertical amplitude, vertical tilt, horizontal amplitude, and horizontal tilt for each of the three colors. Today, reds and greens (top, bottom, right side, left side) work together, while the blue controls are separate potentiometers or a coil on the bottom. Because of these red-green groupings and blue independents, dynamic convergence is considerably easier with the new system and can be done in much less time.

The principle of both systems, nonetheless, is the same. The vertical and horizontal output tubes or transistors supply a sawtooth voltage that translates into a parabolic current in the appropriate dynamic convergence network. These coils are resistive as well as inductive, however, and the sawtooth voltage must sometimes be modified (by a diode) to produce a shaped parabolic current. Probably the best example to consider is the RCA CTC 40-44 dynamic convergence operation.

Vertical Convergence

The vertical convergence circuitry generates the required parabolic current waveshape by shaping a partially integrated vertical sawtooth voltage. A simplified schematic is shown in Fig. 4-10.

The vertical sawtooth voltage is developed across R126, a 39-ohm resistor connected in series with the vertical output transformer primary and B+. The vertical convergence circuit directs current from this sawtooth voltage source through two paths. The path used to converge the upper half of the picture exists through C802, differential resistors R804 and R805, the red and green vertical convergence coils, diode CR803, to B+. Capacitor C802 and resistors R802 and 803 form a shaping circuit in which the values are chosen for optimum convergence. The other path is used to converge the lower half of the picture. During this time, diode CR803 is no longer conducting and the current flows through R803, differential resistor R805, the convergence coils, amplitude resistor R808, the shaping network CR801, CR802, R801 and capacitor C801. The amplitude controls, R803 and R808, provide control over the amount of correction of vertical lines, while differential controls R804 and R805 correct the error in horizontal lines at the top and bottom of the scan.

The blue vertical convergence circuitry, simplified in Fig. 4-11, provides for vertical convergence of the blue horizontal lines as the vertical scan progresses. The correction current required is provided by the same voltage source used to converge the upper and lower halves of the picture in the case of red-green. The blue circuitry provides correction of either polarity as well as of variable magnitude.

Horizontal Convergence

A simplified schematic of the horizontal convergence circuitry is shown in Fig. 4-12. Pulses at a horizontal rate are integrated by L803, R816, R817, C807 and C808, forming a sawtooth voltage waveform which is applied to the red and green horizontal convergence coil windings. This sawtooth voltage waveshape creates the necessary parabolic current waveform in the coils. Clamping circuits, consisting of CR805, CR806 and associated resistors, act to rectify a portion of the circulating current, thereby adding a DC component to the convergence current waveform. This DC component insures that convergence coil current is substantially zero when the beam is in the center of its scan, thus easing setup procedures.

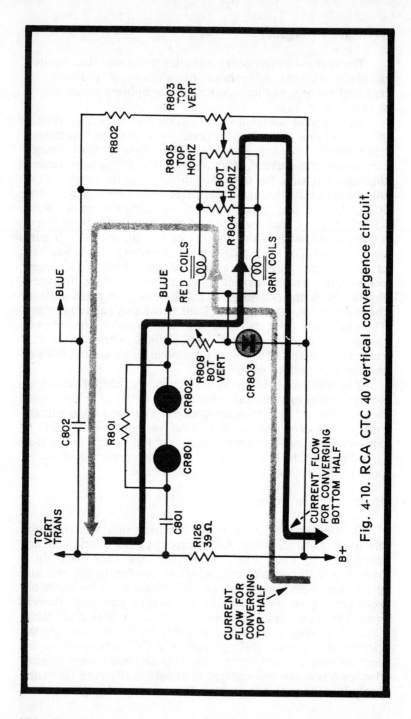

Fig. 4-10. RCA CTC 40 vertical convergence circuit.

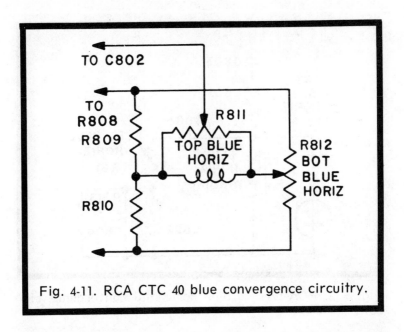

Fig. 4-11. RCA CTC 40 blue convergence circuitry.

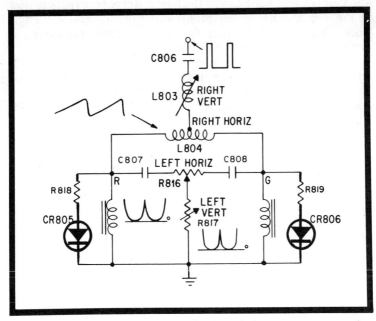

Fig. 4-12. RCA CTC 40 R-G horizontal convergence circuit.

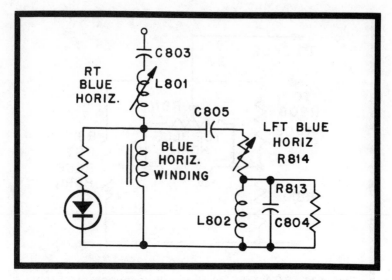

Fig. 4-13. RCA CTC 40 blue horizontal convergence circuit.

Components L804 and R816 allow different currents to flow in the red and green horizontal convergence coil windings. This "differential" current flow is to correct for errors in horizontal lines on the left and right sides of the scan.

The blue horizontal convergence circuitry (Fig. 4-13) operates on the same principle as the red and green horizontal convergence circuitry just described. However, a waveshaping network, consisting of L802, C804, and R813, adds a second-harmonic sine wave of the proper amplitude to the blue convergence coil current. The resulting waveshape optimizes the blue convergence.

Fig. 4-14 is the complete convergence board schematic, showing the entire circuit, with red, blue, and green electrodynamic convergence coils at the bottom outside the heavy dotted line. The "balloon" numbers, 58 through 68, refer to the waveforms, which indicate the p-p voltages, since the current in each case is very small.

The CTC 46 version of this convergence assembly (Fig. 4-15) has some changes in the top and bottom R-G horizontal and vertical adjustment sections, principally in component arrangements, plus the addition of almost half again as many diodes. This seems to mean more control over the convergence circuits, with less interaction among the various adjustments. Diodes will shunt, guide, block, and rectify AC

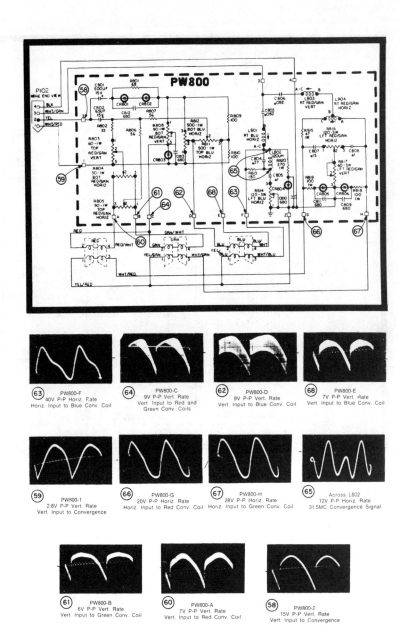

Fig. 4-14. Complete dynamic convergence board schematic with the convergence yoke assembly coils on the bottom (CTC40-44). Refer to the waveforms for drive and resulting circuit characteristics.

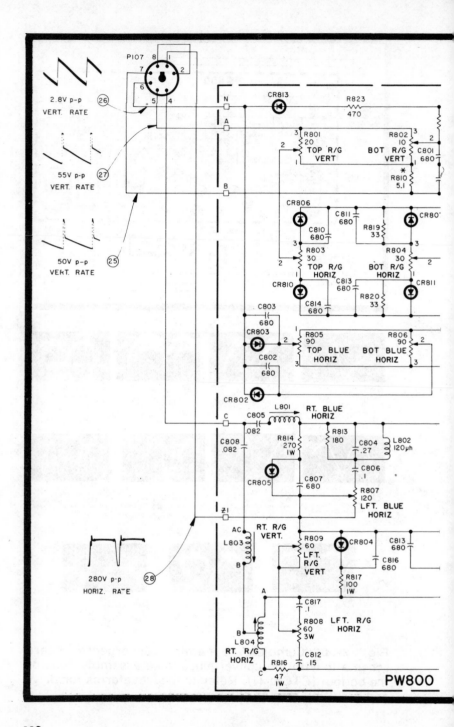

2.8V p-p
VERT. RATE
(26)

55V p-p
(27)
VERT. RATE

50V p-p
(25)
VERT. RATE

280V p-p
(28)
HORIZ. RATE

PW800

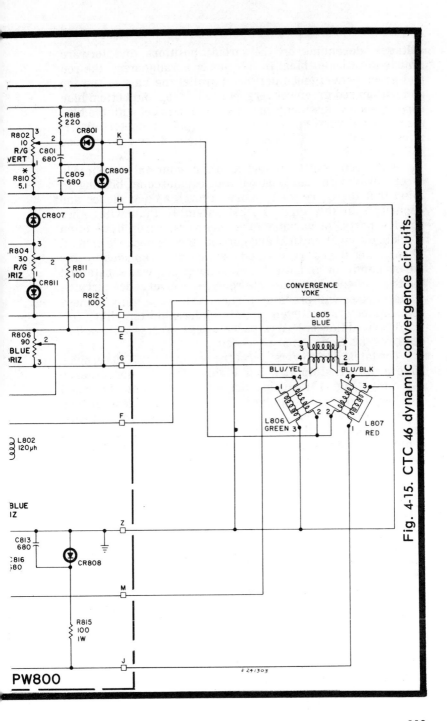

Fig. 4-15. CTC 46 dynamic convergence circuits.

PW800

E 241303

119

voltages, depending on the circuit positions and forward (anode-to-cathode) bias. In this newer arrangement, the red and green convergence portions parallel the blue in the top and bottom red-green convergence, while top and bottom blue convergence adjustments are simply in parallel with the blue convergence yoke coils. The various potentiometers determine current flow into the several portions of the convergence coil windings.

When troubleshooting such a circuit, your only recourse is careful waveform analysis with an oscilloscope, bearing in mind that there are many more failures among diodes and capacitors in this type of PCB assembly than with other passive parts. In vacuum tube receivers, cores have often stuck in the right vertical and horizontal coils because of heat. Then, either the core was freed and probably replaced by any means handy, or an entirely new tunable coil was substituted for the bad one. In solid-state receiver convergence circuits, diode failures will probably predominate, especially if these semiconductors are even slightly underrated for either forward current flow or PIV (allowable peak inverse voltage that destroys the diode when exceeded).

The latest Zenith versions of the original 4B25C19 receiver have a rather simplified but highly effective convergence assembly (Fig. 4-16), using but four diodes and the usual number of coils and potentiometers in the same general type of arrangement for dynamic convergence adjustments. Convergence assembly iron core series coils are shown in the center of the diagram, with the 9-pin connecting plug on the top right.

Two-Pole Convergence

Some small screen receivers (Fig. 4-17) have a 2-pole convergence system where the blue controls act as in the standard 3-pole system. There is a red-green lateral magnet that moves the red and green beams in opposite directions until the two beams cross. At this point, you have RG convergence and the correct control setting. The red-green horizontal and the red-green vertical balance controls are added to balance both magnetic fields for top and bottom and right-left sides so that each is supplied equal parabolic waveshape correction. This is especially true at the beginning and ending of the vertical and horizontal sweeps. The amplitude controls, then, are simply the amount of correction applied, while tilt is the placement of correction applied.

In setting up and converging the average receiver, the new ones are rarely touched unless there has been meddling or a new cathode ray tube has to be installed. Occasionally on new products, the factories have "people problems," but this is normally corrected promptly after a new chassis' initial run.

QUESTIONS

1. What makes color fringing?
2. Color impurities are caused by _____ ?
3. Deflection yokes are positioned where at the start of purity adjustments?
4. Do you adjust static convergence initially before or after purity adjustments?
5. Do you use a dot generator before or after **initial** purity adjustments?
6. Why do you degauss a receiver?
7. When you spread the purity tabs, the magnetic field strength decreases, true or false?

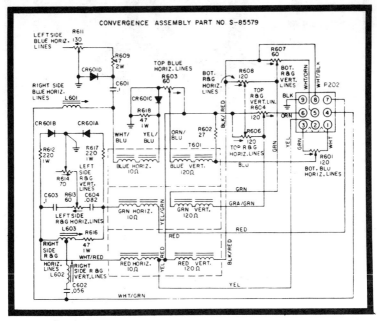

Fig. 4-16. Zenith 4B25C19 series dynamic convergence board schematic.

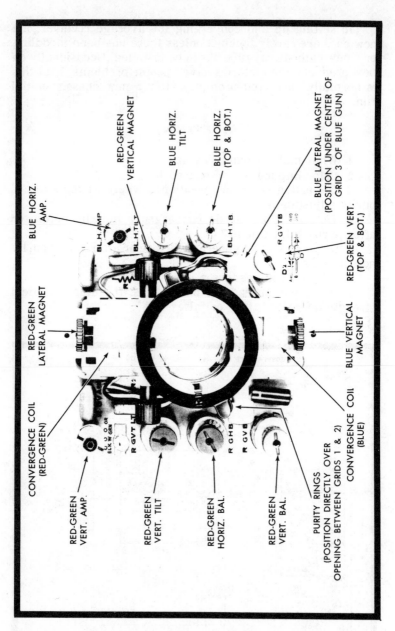

Fig. 4-17. View of the two-pole static and dynamic convergence components on the CRT neck in Zenith's 14CC14 through '16 chassis. Notice the red-green lateral magnet and different control terminology.

8. What percentage of new color sets have automatic degaussing?

9. In gray-scale adjustment, your first adjustment is selecting _____?

10. Vertical convergence circuitry generates what kind of current waveshape?

11. Horizontal convergences form a _____ waveform?

12. What type convergence assemblies are found in small-screen receivers?

Chapter 5

Tuners, Video IFs & AFTs

Since many of these circuits were initially discussed in Chapter 1, it is necessary to cover here only those portions that have specific significance, with additional emphasis on the evolution of signal processing by vacuum tubes, discrete transistors and integrated circuits.

Fig. 5-1 is a system block diagram in which the UHF tuner output is plugged directly into the VHF tuner, where sections of the lower band tuner function as additional RF amplifiers to boost Channels 14 through 83 and pass them at IF frequencies to the video IFs. Of course, if the local UHF or VHF oscillator is off-frequency and the automatic fine tuning (AFT) circuit is engaged, the AFT (within limits) puts the local oscillator(s) back on frequency so that the 42.17-MHz chroma and 45.75-MHz video carriers are both somewhere near the mid (half power)-points on opposite sides of the response curve (Fig. 5-2). The typical IF response has a flat top and sloping sides, depending on trapping and the design characteristics of the various IF stages.

You will notice that we show three IF amplifiers in Fig. 5-1. Some tube receivers have two IFs and some transistorized sets have four IFs, but most receivers have a 3-tube or transistor IF system. Most IFs are stagger tuned (each circuit tuned to a slightly different frequency) to achieve the desired gain and bandpass.

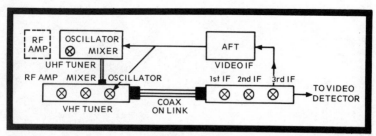

Fig. 5-1. Familiar block diagram of a tuner, AFT and IF system.

It is **imperative** you know and understand video IF alignment waveforms completely if you expect to deal effectively with color television. A "sloppy Joe" monochrome response may satisfy uncritical mothers with cataracts, but the average wage-earner is going to expect a substantial return for his $600 to $800 investment (if he buys a top-rated receiver).

TUNERS

The VHF tuner simply selects any one of 12 channels between 54 and 216 MHz. The local oscillator operates 45.75 MHz above the video carrier, and this results in a center 44.25-MHz beat difference that becomes the mid-video IF frequency.

The response curve of any tuner, of course, is very similar in appearance to the vacuum tube IF response, with the sound and video carriers separated on the top by the standard 4.5-MHz difference (the sound carrier is always the higher frequency). At Channel 2, the picture carrier is 55.25 MHz, while the sound carrier is 59.75 MHz. On Channel 13, the picture carrier is 211.25 MHz and the sound carrier 215.75 MHz. This applies equally to the UHF spectrum where the 4.5-MHz frequency difference remains the same.

Vacuum Tube Switch Tuners

The RCA KRK 120T UHF tuner (bottom portion) Fig. 5-3, uses a semiconductor, but then all UHF tuners have been transistorized for a number of years. The UHF input passes through a 300-ohm centertapped balanced coil. The desired signal is selected by reactive gang-tuned capacitors C2, C4 and

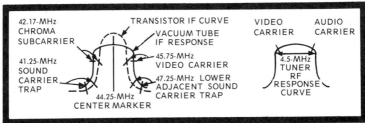

Fig. 5-2. Typical tube and transistor IF and RF response curves (they're often similar), showing chroma and video carrier positions, plus two basic trap frequencies. Heavy line is the vacuum tube receiver curve, while the dashed line is a semiconductor receiver IF waveform. Center frequency for all IF waveforms is 44.25 MHz.

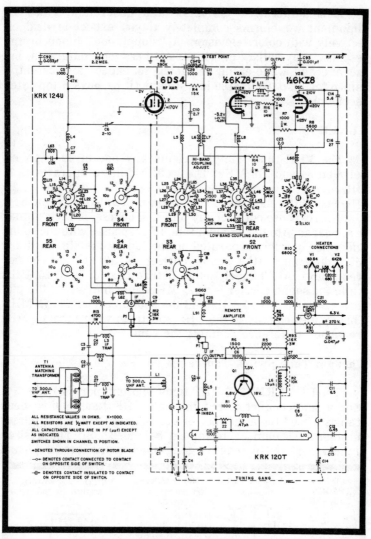

Fig. 5-3. Schematic of RCA's KRK 124U, W-KRK120T tuner used in CTC25 series receivers.

C14 that resonate with coils L2, L3 and L8. C1, C3, and C13 are variable tuner trimmers. Coils L4 and L10 are inductive couplers for oscillator Q1 and mixer diode CR1, while L6, shunted by R2, is part of the damped load, along with R93 from the power supply. R5 and R6 are base biasing resistors (voltage dividers) for Q1 which, again, oscillates at 45.75 MHz above the video carriers in the 470- to 890-MHz UHF band. Diode CR1 mixes the incoming RF with the oscillator signal. The beat difference is slightly filtered by C5 and passed to the Channel 1 position of the detented VHF tuner.

The **KRK124U VHF tuner** accepts this IF signal and the V1 ceramic 6DS4 RF amplifier becomes a UHF IF amplifier along with the 6KZ8 mixer. The VHF oscillator is cutoff from B+ in the Channel 1 UHF position. Since the UHF IF output already has some slight filtering, IF feedback would be pretty difficult, and the tuner sees only frequencies at 470 MHz and above. Heavy IF filtering and moderate high-pass filtering is hardly needed.

When the UHF tuner is not used and the VHF tuner is fully supplied with input signals and DC operating voltages, the input circuit and accompanying filtering become quite important (Fig. 5-4). VHF signals are applied through the an-

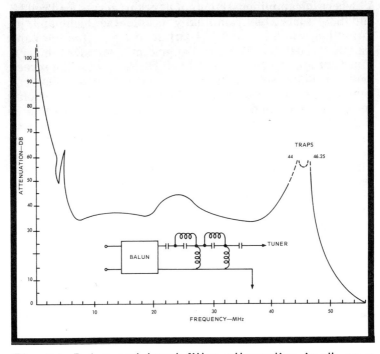

Fig. 5-4. Balun and input filter attenuation in db as a function of frequency in MHz. Notice that undesirable signal rejection from 0 to 50 MHz is no less than 32 db and more than 50 db at the traps. (Courtesy Dave Carlson, RCA)

127

tenna matching transformer (T1), called a balun (balanced-to-unbalanced), that transforms the 300-ohm antenna and transmission line impedance to a 75-ohm input for the RF amplifier. C1, L1, C4 and L3 form tuned IF traps that prevent IF frequencies from radiating outward from the receiver to the antenna and also block frequencies around 40-plus MHz and below from entering the tuner. At the same time, high-pass T filter (C2, C3, and L2) also aids in suppressing incoming signals below 54 MHz. (An 88- to 108-MHz FM trap might be something of a variable notch-type filter for specific passband rejection.) A graph of balun and high-pass filtering is shown in Fig. 5-4. Minimum rejection is 32 db at 50 MHz.

The signal input now passes through S4 and S5 to the series switched inductances on the S5 front wafer, which by shorting action selects the correct reactances and time constants so that the RF amplifier sees only the 6-MHz band of frequencies to which the channel selector is tuned. The signal is passed to V1 where the amplifier RF output develops across load coil L5 and S3 (front). L6-L7 mutually couple the high-band in-formation, while L33 and L44 adjust and pass the low-band signals through S2 (front) to the grid of V2A (6KZ8 mixer). This tube uses a common +270-volt DC supply for the screen grid and plate, separated and decoupled by two coils (one peaking) and a pair of current-limiting and voltage-dropping resistors. The tuned local oscillator, as stated, operates at 45.75 MHz above the video carrier, and this signal is also put to the mixer grid through L36, L37 or S2 (rear) and L8. The beat difference is the 45-MHz IF video carrier that is now passed through IF output J2 to the IF amplifiers. The DC supply reaches the local VHF oscillator through S1 and L60, an AC blocking reactance. Additional switch coils determine the frequency (or repetition rate) of the local oscillator (V2B) in conjunction with C14, C16 and the associated resistors.

Transistorized Turret Tuners

Although the basic function of any tuner is to select the desired RF frequency and convert it into separate video and audio carriers, the transition from tube to solid-state tuners has been rather long and sometimes painful. All sorts of common-emitter, common-base, and cascode RF con-figurations have been tried, some with good, mediocre and poor results and others not worth using, until rather recently. The story is pretty well told in Figs. 5-5 through 5-8.

Any tuner must have certain creditable performance characteristics that include satisfactory low noise figures, adequate 6-MHz bandpass with good adjacent-channel rejection, a relatively broad fine tuning range, little drift, tolerable input impedances, low standing wave ratios (SWR), limited cross modulation (there is always some), good gain, mechanical ruggedness, ease of servicing, acceptable low local oscillator radiation and noise-preventive shielding.

The most difficult design part of any tuner is the RF amplifier, where gain, noise, SWR and cross modulation are all wrapped up in a single (or cascode) device that determines much of the overall characteristics of the tuner. Fig. 5-5 amply illustrates the effects of cross modulation (where two strong signals force the generation of additional RF signals by modulating one on the other) in terms of gain reduction for at least the unproductive beginning with the unwanted sum and difference beats. The graph specifies the percentage test figure as 1 percent.

The A466 common-emitter bipolar begins cross modulation at 35 millivolts and climbs steadily, with a gain reduction up to 25 or so db to 700 millivolts (0.7 volt). Without taking into account other factors, the raw standing wave ratio

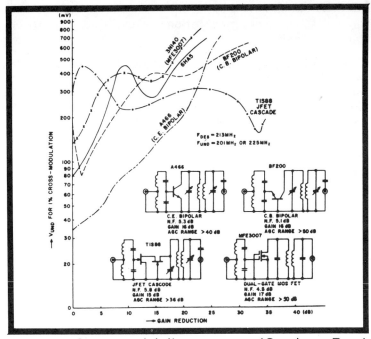

Fig. 5-5. Cross modulation curves. (Courtesy Frank Hadrick, Zenith)

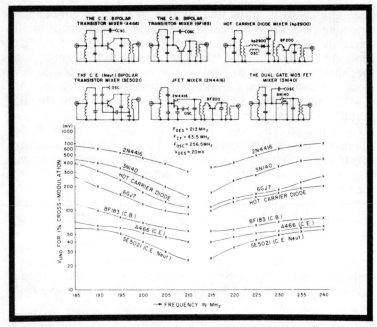

Fig. 5-6. Mixer cross modulation curves. (Courtesy Frank Hadrick, Zenith)

is Emax divided by Emin, or 700 divided by 35—a rather undesirable SWR of 20. The BF200 common-base bipolar looks somewhat better, since cross modulation doesn't begin until about 180 millivolts, dips to 80, but still extends to 600 at the 25 db down point—a more modest SWR of 8.75.

Next, a junction field-effect (TIS88) transistor was tried and 1-percent cross modulation began at 300 millivolts, full

PERFORMANCE COMPARISON CHART				
	TYPICAL FET GVG TUNER		TYPICAL TUBE GVG TUNER	
	CH. 2-6	CH. 7-13	CH. 2-6	CH. 7-13
N. F. (dB)	<5.3	<5.5	<5.3	<6.5
GAIN (dB)	>36	>32	>38	>33
IMAGE (dB) REJECTION	>77	>62	>75	>55
I. F. (dB) REJECTION	>74	>85	>70	>75
GAIN (dB) REDUCTION	>58	>50	>50	>49
VSWR	<3.5	<3.0	<4.7	<3.5
SPURIOUS (dB) REJECTION	>88	>75	>80	>50

Fig. 5-7. Performance comparison chart with composite data.

gain, then increased to 150 mv at about 2 db reduction, and fell thereafter down to about 150 mv at 35 db gain reduction. This raw SWR amounts to a 3:1 figure over a much wider range, probably much more than is needed.

The final amplifier tried was a dual-gate metal oxide semiconductor that produced a gain of 17 db with an AGC range of better than 50 db—a rather marked improvement over even the JFET, and better than the common-base BF200 bipolar transistor. The MOSFET MFE3007 began 1-percent cross modulation at 150 millivolts and increased to about 700 at the 25 db gain-loss point, an overall SWR in this case of 4.67:1. Observe how this compares with the curve for the 6HA5 tube that starts at 80 millivolts and finishes about 650.

The JFET looks good, but even in the **cascode** (drain tied to source) configuration, its noise factor is 5.8 (the highest), its gain is less, and AGC range is limited to about 36 db. So the dual gate MOSFET, obviously, wins with a low noise figure of 4.5 db, a higher gain, and a variable AGC tolerance of more than 50 db. Even better figures than the hot, current-consuming 6HA5 vacuum tube that has been an industry standard for some time.

Next to be developed was Zenith's mixer, and Fig. 5-6 shows the 1-percent cross modulation figures, in millivolts as usual, but this time as a function of frequency in MHz. The tube comparison was a 6GJ7, another good performer. Test frequencies began at 185 MHz and continued to 240 MHz, since the mixer operates at these frequencies that are apparently more critical in the upper channels. At any rate, the same types of configurations as used in the RF amplifier tests were tried, including hot carrier diodes, yet the results proved pretty much the same. The 2N4416 JFET was good, but overall performance was better with the dual gate MOSFET and, as you can see by comparison, it performed better than the 6GJ7 standard vacuum tube mixer. So in digital as well as analog (linear) applications, MOSFETs and CMOS (complementary MOSFETs) are now becoming the leaders in many types of signal processing, especially since most now have protection against electrostatic gate punch through.

A final comparison of FET and tube-type tuners is shown in Fig. 5-7, beginning with noise figures and ending with unwanted frequency rejection. Arrows pointing toward a number designate "greater than" and arrows in the opposite direction mean "less than." As you can see, only in db gain does the tube type have any small advantage over the FET, and those differences are minuscule.

The completed unit is shown in Fig. 5-8, and has been designated Zenith's 175-1800 Series solid-state tuner. The three

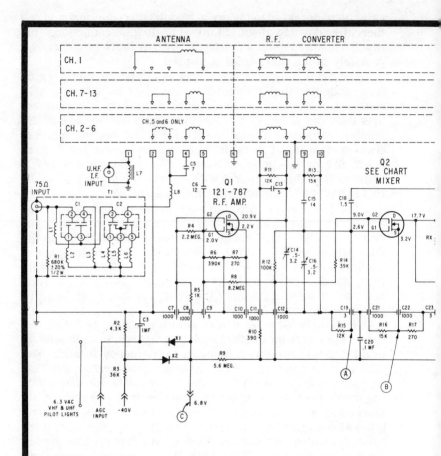

Fig. 5-8. Zenith's 175-1800 solid-state Gold Video Guard VHF turret tuner with two MOSFETs and a bipolar oscillator.

strips at the top of the schematic represent the actual rotating coils, where Channel 1 is the UHF input and Channels 7-13 and 2-6 are the RF-antenna signal receiving portions. Other coils tune the converter and oscillator so that the entire tuner responds to the incoming 54 to 216 MHz frequencies.

Signals come in through a 75-ohm input and across substantial traps, through the antenna coils to the G1 gate, which is slightly biased by 2.2-megohm R4 and AGC controlled through guide diode X1. Q1 amplifies the incoming microvolt-level signals under varying delayed AGC control, depending on signal strength, and puts out a voltage through Terminal 8,

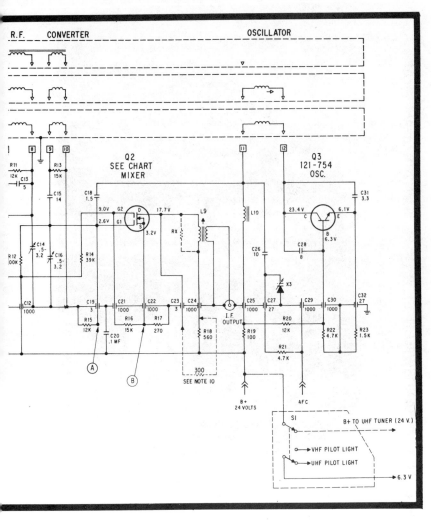

which is inductively coupled to gate 1 of the mixer (Q2). Again, G2 of this MOSFET is the more strongly biased electrode, while G1 receives both the amplified RF signal and the signal from the oscillator, which is applied to Q2 through the coil coupling between Terminals 11 and 12. Automatic fine tuning (when operating) controls the oscillator through varactor X3 when the video IF 45.75-MHz frequency varies from its midpoint position on the slope of the IF response curve. This process is repeated from channel to channel as the turret is rotated, inserting various inductive tuning and coupling for the three stages. Notice the many feed-through

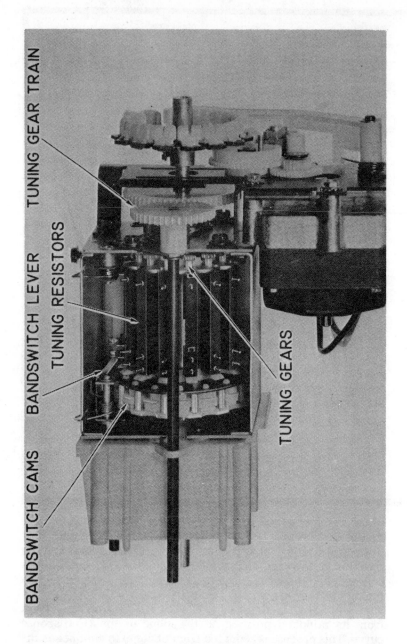

Fig. 5-9. UHF-VHF bandswitch, cams, and tuning resistors in Zenith's all-varactor tuner.

capacitances along the common line of the tuner's lower "ground" edge. They are used both as filters and AC grounds and are quite effective (C7 through C12, for instance).

A UHF Varactor Tuner

The varactor is a special back biased diode that operates like a capacitor, except that it is entirely DC voltage controlled...in this instance between 0.5 to +28 volts from a regulated supply source and energized by a carefully controlled voltage drop across a precision resistor. The tuning resistors, levers, and gear train, shown in Fig. 5-9, are part of Zenith's varactor VHF-UHF tuners featured in the deluxe receivers. Each tuner resistor is separately adjustable (programmed), and when the tuner is rotated to an individual channel, a discrete voltage develops across one of the resistors and tells the diode(s) what capacitance it should represent, thereby tuning the RF amplifier and local oscillator. The VHF tuner, of course, operates on the same general principle, and could well be included here, except that we discuss next a special printed-circuit tuner built along the same lines.

This tuner (Fig. 5-10) can accept either 72- or 300-ohm inputs, with a balun mounted on the antenna terminal board itself, rather than on the tuner as has been done in the past. A second innovation is that J102 is both the 75-ohm VHF and UHF antenna input. The VHF is separated from the UHF signal by C101 and L101, which operate as a low-pass filter, while C102 couples the UHF directly into the Q101 RF amplifier tuning inductor.

Since the UHF tuner has neither low nor high bands, but a single 470 to 890 MHz frequency range, programmed voltages to varactors CR101, CR102, CR103 and CR104, in conjunction with inductors L102, L104, L108 and L110 control both the RF amplifier channel selection and the repetition rate of the common-base oscillator. RF and oscillator signals are passed through mixer diode CR105 and L110 to the base of the common emitter amplifier, Q103, which feeds the IF to the VHF tuner. Tunable capacitors CX101 (high-end-of-band tracking) through CX103 are additional tuning resonances of the manual variety. The same is true of L103 (low-end-of-band tracking), L106, L109, and L111. L112 is fixed frequency only.

Printed-Circuit Varactor VHF Tuner

This tuner was developed by Motorola Semiconductor, Phoenix (Ben Scott, Applications), to demonstrate the use of

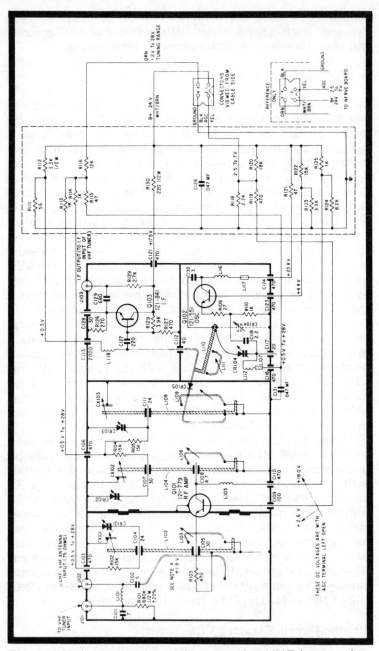

Fig. 5-10. Schematic of Zenith's varactor UHF tuner using a grounded-base AGC-controlled RF bipolar input.

their MV3102, MV3141 varactor diodes and MPN3401, MPN-3402 pin switching diodes. The tuner has both a high and a low band, with diodes D2 (five of them) either being reversed or forward biased. The diode bias level and inductors L1, L4, L7 through L10 change the frequency response of the tuned circuits. Thereafter, the D1 series of varactors are programmed by tunable resistors.

Again, a MOSFET is used as the RF amplifier because the tuner noise figure, a low degree of cross modulation, power gain, small VSWR, good AGC range, and relative changes of input impedance with AGC all depend on it. Here Motorola uses its MPF121, which has both low input and output capacitance, a common-source typical noise figure of 2.6 db, and a **minimum** common-source power gain of 17.

The mixer is a bipolar cascode type with the first collector DC coupled to the second emitter and a full 24 volts DC across both transistors. The first transistor circuit, obviously, is a common-emitter type, while the second is common-base. This configuration was used, because of the extra gain it produces, and the excellent isolation provided between input and output.

The oscillator must have a stable, strong output between 101 MHz and 257 MHz, and the selection of the tuning diode is critical because the oscillator may swing the diode into forward conduction with small bias voltages and radically change frequency. An MPS-H11 is used as the oscillator. Since the varactor diode's capacitance determines many of the passive RLC component values, a designer has to limit himself to the smallest reliable tuning voltage. This will lower the Q at small bias voltages and reduce the possibility of intermodulation distortion where two different frequency sine waves are injected and the power output is measured with no clipping. Fig. 5-12 lists Motorola's findings of this tuner's performance per channel.

The foregoing should give you a comprehensive introduction to the various types of tuners, their electrical and mechanical features, progress over the last several years, and some of the stiff design hurdles that engineering must always overcome. All these factors are good general lessons that can contribute considerably to the overall understanding of any tuner's functions and desirable characteristics.

VIDEO IFs

The function of these variously numbered stages (although three is now typical) is to accept center frequency

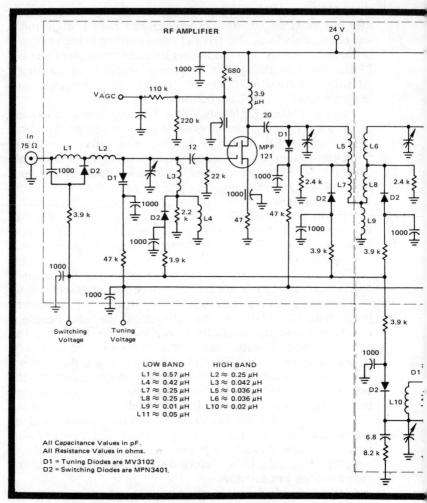

Fig. 5-11. Schematic of the Motorola printed circuit tuner using varactors, and pin diodes.

signals of approximately 44.25 MHz, which contain the video carrier at 45.75 MHz, audio carrier at 41.25 MHz, and the chroma carrier at 42.17 MHz, and process these signals with sufficient gain and bandpass.

How these IF strips are designed, shielded, controlled and supplied determines much of the operating efficiency of the remainder of the set. Luminance and sync information should fill the AM passband up to about 3.5 MHz. FM audio is derived from the 4.5 MHz heterodyne heat between the eventually

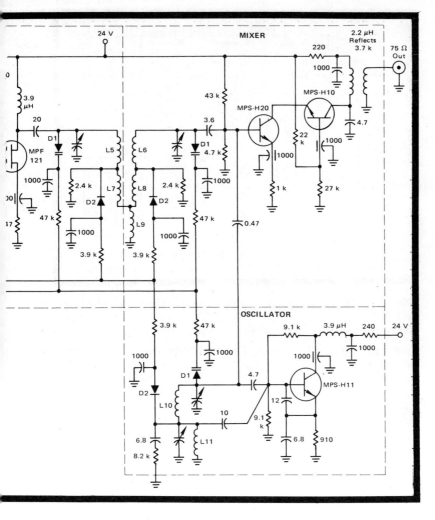

trapped 41.25-MHz audio carrier and the 45.75-MHz video carrier. Chroma is clustered at the 0.5-MHz sum and difference frequencies about the suppressed chroma subcarrier at 3.6 MHz. Video IF strips often develop a tilt that must be compensated by a reverse, equalizing tilt in the chroma amplifiers.

Meanwhile, have you often wondered what happened to the wondrous 4.2-MHz video bandpass that some writers still mention? For many years, such broad response hasn't existed. You may think you're seeing something like it, but appearances are deceiving. The practical limit to video

Channel	Tuning Voltage (V)	Power Gain (dB)	Noise Figure (dB)	VSWR
2	4.3	31.7	5.5	2.1
3	6.7	32.5	5.5	2.1
4	9.4	33.0	5.5	2.7
5	16.0	33.0	5.7	3.0
6	24.0	33.0	6.0	2.0
7	10.9	32.8	4.5	2.0
8	12.3	33.5	4.3	2.4
9	13.8	34.0	4.5	1.9
10	15.7	34.5	4.3	1.9
11	17.7	34.5	4.4	1.6
12	20.1	35.0	4.4	1.6
13	23.0	35.7	4.5	1.2

Fig. 5-12. Printed-circuit tuner performance per channel.

bandpass today is little more than 3.5 MHz. On a monochrome receiver, for instance, if the video response exceeded this bandwidth, color around the 3.6-MHz carrier frequency would immediately appear as a disagreeable herringbone background interference that most people wouldn't tolerate. On the other hand, if a color receiver went much beyond 3.5 MHz, the clash between color and monochrome luminance intelligence would actually cause chroma **desaturation**, and you'd have washed-out colors. So bandwidth and picture resolution are really a series of manufacturing compromises that produce the "best overall picture."

Before we get into the actual IF discussion, the trap and carrier frequencies in Table 5-1 should be helpful in understanding why certain circuits are designed the way they are. Almost all color receivers have at least seven of these eight IF traps and sideband-carriers, while Motorola's Original Quasar has the eighth—the 35.25-MHz upper-adjacent sound trap. You would do well to commit these important frequencies to memory. Remember, the video-chroma carriers and sidebands all appear on the sides of any swept IF response curve, while the traps determine rolloff characteristics and are all situated along the baseline, just like the 41.25-MHz audio-carrier trap shown in Fig. 5-11.

Carriers (MHz) and Sidebands	Traps (MHz)
41.67-Lower chroma sideband	35.25-Upper adjacent sound trap
42.17-Chroma subcarrier	39.75-Upper adjacent pix. carrier
42.67-Upper chroma sideband	41.25-Sound carrier
45.75-Video carrier	47.25-Lower adjacent sound carrier

Table 5-1. IF trap and carrier frequencies.

Vacuum Tube Video IFs

In Fig. 5-14, coming through P101 are the video IF and sound IF signals, where the latter is partially trapped by lower adjacent-sound carrier trap L301-C302, C303, and sound reject adjust R301. Video information and 41.25-MHz audio are then passed through the impedance matching and coupling first IF transformer to the grid of the 6JH6 first IF. Here the DC voltage will be 0 if there are no signals and somewhat negative if there are incoming signals. L302, C304, L308 and C310 are LC filament filters. The AGC control voltage is applied to the first and second IF stages through the transformer secondaries. The AGC voltage arrives at PW300 terminal E.

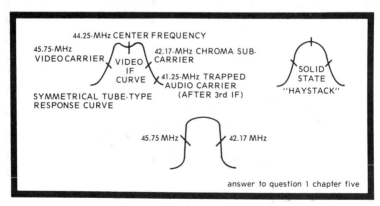

Fig. 5-13. Solid-state video IF curve shown side by side with tube IF curve, indicating center IF, chroma, video, and sound markers.

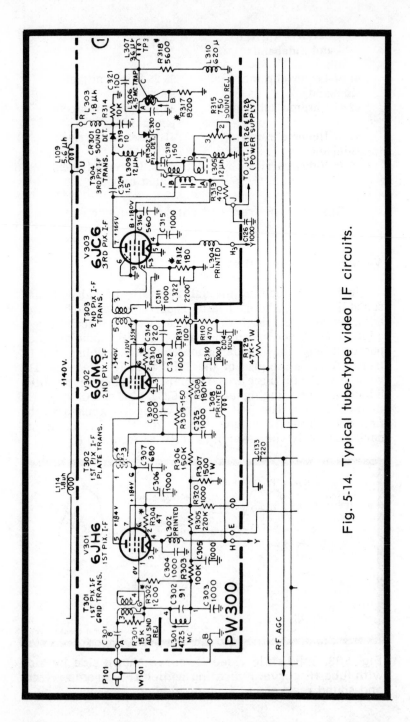

Fig. 5-14. Typical tube-type video IF circuits.

The 6JH6 is a semi-remote cutoff pentode used in gain-controlled video IF stages and is an offspring of the old 6BZ6. Open transconductance is 8,000 micromhos, but with a control grid voltage of -4.5 volts, it drops to between 400 and 900 micromhos.

Terminal F is the 405-volt B+ supply point and there follows a series of dropping resistors and filter capacitors from the output of the second IF, straight through to the grid of the first IF, including resistors R308, R306, R305, etc., with dividers R307, R311 and others. And although T301, T302, and T303 show no internal cores or variable arrows, you can be sure at least one section can be tuned. The second IF amplifier (6GM6) is another semi-remote cutoff pentode with a trans-conductance of 13,000 micromhos. It's cathode is at virtually the same potential as V301's plate, so V302 is called a "stacked" amplifier and, because of its high cathode voltage, the plate voltage can go to 340 volts (340-184 or 156), and deliver a very high output swing with reference to ground.

Signals out of V302 are transformer (T303) coupled into the grid of V303, a sharp cutoff pentode with a high trans-conductance at low B+ voltages. The cathode of this tube is referenced directly to ground through a 180-ohm resistor bypassed by C315 to remove signal degeneration and permit maximum DC current flow plus AC tube operation. Bypasses for the screens and suppressors of these tubes also prevent extraneous signals from forming on the electrodes, which would cause undesirable tube degeneration.

T304, following V303, is both the third IF transformer and the 41.25-MHz sound carrier trap. R315 is a second sound reject potentiometer, shunting the reactive impedance of L305, a 12 microhenry peaking coil. L309 is another coil, but this time it is called an RF choke since it is just before the CR301 sound detector. CR301 picks up the sound on a new FM beat carrier of 4.5 MHz, the difference between the 45.75-MHz video carrier and the 41.25-MHz sound carrier. From the sound detector, the 4.5-MHz signal goes to the sound IF.

As a further precaution that no sound reaches the video circuits, a 4.5-MHz sound trap (L306) follows video detector CR302. With the video and sync signals ready for further processing, information now goes through peaking coil L307 to the succeeding circuits for sync, chroma, and video (luminance) separation.

TRANSISTOR VIDEO IF STRIP

From the world of vacuum tubes to that of transistors isn't such an enormous transition as one might expect (Fig. 5-15).

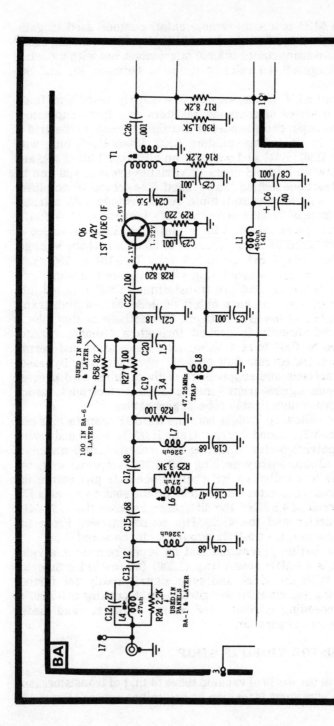

144

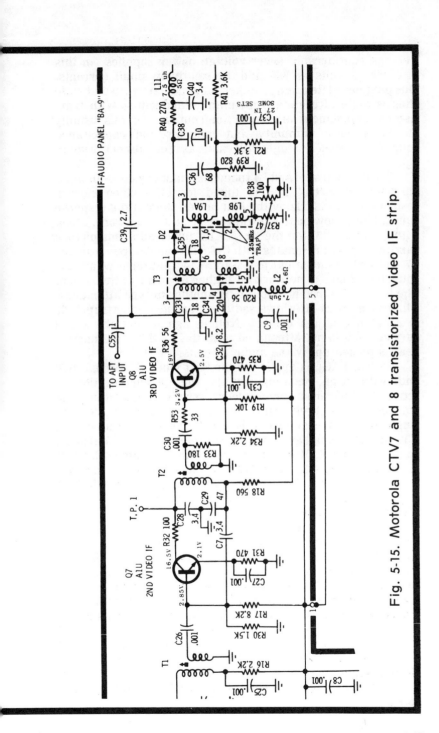

Fig. 5-15. Motorola CTV7 and 8 transistorized video IF strip.

Yet, the considerably lower voltage power supplies (in this instance, 20 volts vs 405) and the relatively small currents, plus plug-board flexibility, makes the modular type solid-state chassis quite attractive from many standpoints. With transistors, impedances are lower, of course, and this usually increases noise immunity, but inductor and capacitance coupling between IF stages still exists as it did almost 20 years ago.

The RLC parallel and L4 tuned resonant circuit is an input bandshaping circuit and is tuned for best video carrier response at 45.75 MHz on the half power point of the response curve. L5 is a low pass aid filter, part of a pi network, along with C13 and C15, for the 39.75-MHz upper-adjacent picture carrier trap that resonates with C16. R25 is a damper resistor. L7 and C18 are identical to L5 and C14, and form what now could be called a pi network for higher frequency attenuation. The L7 and C18 trap is followed by another pi-type attenuator, which is the 47.25-MHz lower-adjacent sound carrier trap; R27 is a balance adjust control. Notice, again, that the sound carrier itself is not attenuated until the 4.5-MHz intercarrier beat is picked up by the audio circuits.

Forward AGC is applied to the first video IF through R28 and, with large incoming signals, the Q6 current flow is reduced with overbias and so decreases the signal output. The emitters of all three transistors are resistively biased for single battery operation and the most stable operating points, R30 and R17 being base divider examples. Here, for instance, you have a 20-volt DC supply. So base bias amounts to E x 2.2K divided by (2.2K + 8.2K) or 4.23 volts, less the voltage dropped from the transistor base to emitter, since the potential difference of 0.75 volts between the base and emitter shows the transistor is conducting.

Capacitors C28 and C29 are stabilizing shunts, and C7 is a little degenerative feedback path that, with C29, is also a capacitive divider. C7 also helps control the base of the second video IF, since it is not AGC controlled. R33 (across the secondary of T2) is an anti-ringing and stabilizing resistor, while L2 (off the lower primary of T3) is a choke to keep IF frequencies from the DC supply. T3 is the third IF transformer with a 41.25-MHz sound trap in its secondary, adjusted by R38 for best trap notch. The automatic fine tuning signal sample and the sound signal are taken from the collector of the third IF just before the primary of double-tuned T3. Diode D2 is the video detector diode, followed by a small filter and peaking coil L11.

Integrated Circuit Video IF

RCA unveiled the first all-integrated circuit video IF amplifier. It was used in the first U.S. 110-degree 18-inch thin picture tube color receiver, the CTC49. In addition to IF amplifiers, the CA3068 has a noise limiter, keyed AGC, buffered AFT output, sound carrier detection, amplification, with zener diode regulation—quite a subsystem for a little 20-pin integrated circuit that can dissipate 600 milliwatts!

Fig. 5-16 is a block diagram of the system with power supply and pass transistor between pins 15 and 18, plus IF transformers at the top and left of the illustration. Signals come in through double-tuned bandpass shaper T1 to the first IF amplifier, which drives the AGC, additional IF stages, video amplifiers and sound channels. The outputs, of course, are sound and video, with separate RF AGC voltage for the tuner. Fig. 5-17 is a photo of the complete module.

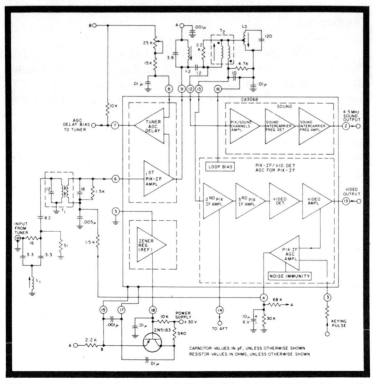

Fig. 5-16. Block diagram of the RCA CA3068 IC video IF, showing internal functions and external components.

Fig. 5-17. Photo of the entire MAK IF module with CA3068 and CA3064 IF and AFT ICs located in the second and third compartments from the left. (Courtesy Tom Bradshaw)

A simplified schematic of the CA3068 appears in Fig. 5-18. Darlington stages Q19 and Q20 drive parallel-connected amplifier transistors Q3 and Q4. With increased input, Q1's conduction is decreased, and Q4 then is brought into operation by the signal input at the Q9 emitter. With Q1 cutting off, less of a hold is applied to the base of zener clamped Q21, and this transistor turns on Q5 for an increased negative output to control the tuner's RF amplifier. Q21 also supplies drive for Q14.

The transformer tuned wideband IF amplifier includes emitter follower Q6, amplifier Q7 and amplifier Q8, with 45.75-MHz signals passing through pin 14 to the automatic fine tuning and full video to the detector. Q8 drives Q22, which is regulated and biased by current source Q9 and the current flowing through the emitter of Q23. R19 and C2 combine as a peak detector with selected time constants that do not introduce differential chroma phase errors nor result in amplitude distortion. Video amplifier Q24 receives DC bias and signals from the video detector and transmits them to amplifiers Q12 and Q25.

Transistor Q12, when in conduction, determines white level for video output terminal 19. To prevent its conduction when there is no video, the emitters of Q10 and Q24 must have the same DC potential and similar currents. When this happens, Q11 absorbs all the current from Q24 and there is no flow

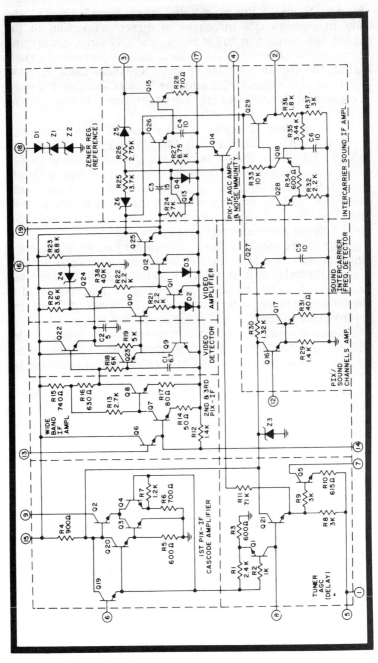

Fig. 5-18. Simplified schematic of RCA's CA3068 video IF.

through D3 or Q12. But when video information is present, the current through Q24 increases in direct proportion to the signal, the current in Q10 is fixed, and Q12 fires, permitting video output at terminal 19. With a further increase in video signal, the DC at the base of Q12 increases, there is a DC decrease at the base of Q25 until the base of Q25 drops toward ground and further signal increases are "bottomed" or clipped.

With a normal signal, Q25 transmits the composite waveform with sync tips keying AGC amplifier Q13, which also affects Q14. When the voltage for Q13's base begins to fall below 0.8 volt, the keying current is diverted increasingly into diode D4. During this time, Q14 discharges an external 10-mfd capacitor, usually connected to terminal 4, at a rate proportional to the base current entering Q14, and this varies the AGC voltage inversely and proportionately.

For noise immunity, the keying current is removed during times of incoming transients or pulses by noise detector Q26, clamp Q15 and pass filter C3 and R27. Q26 and C4 form another peak detector and, when the DC across C4 is proportional to the incoming noise pulse, it turns on Q15, clamping the keying information to ground and shutting off the noise.

Sound is applied to terminal 12 and follower-amplifier Q16-Q17, while detector Q27 and C5 operate on the 4.5-MHz sound difference beat between audio and video carriers and supply this information to Q28. Q28 and Q18 form a differential pair to amplify the sound signal for Q29, the output emitter follower and driver for the sound demodulator.

AUTOMATIC FINE TUNING (AFT OR AFC)

Automatic fine tuning circuits all operate pretty much the same way, although design can vary as you will see. The purpose of AFT or AFC is to detect any variation of the 45.75-MHz video carrier for some 50 or so kHz on either side of center frequency, convert the AC variation into DC, and deliver a correction signal to the UHF-VHF tuner oscillators to bring them back on the selected channel frequency. The AFT isn't normally a difficult circuit to service nor keep on frequency, and reported field service for those units hasn't been at all excessive. "Eyeball" alignment can often be done in the home by simply using a broadcast color signal, manually adjusting the tuner for the correct center frequency, then switching the AFC or AFT in and out of operation while making the transformer adjustments. Operationally defective units, of course, need initial repairs, then actual sweep alignment with a crystal-controlled marker generator to be sure. The operating frequency sample for conventional AFT systems is taken off the third video IF for maximum amplitude prior to demodulation.

Motorola's Discrete AFT

Motorola, with the first all transistor color receiver in 1967, offers a very simple automatic fine tuning circuit that is quite effective (Fig. 5-19). A limiter-emitter follower picks up a signal sample from the third IF through a single-battery biased current amplifier Q1. The Q1 output is fed into amplifier-discriminator Q2. Acceptable frequencies pass into the tuned primary and secondaries of AFT discriminator transformer T1, a circuit that is reminiscent in many respects of the ancient but excellent Foster-Seely audio discriminator. T1 receives an incoming signal equal to the tuned frequency of its secondary, diodes D1 and D2 produce balanced currents that cancel. If the input signal varies above or below 45.75 MHz, a voltage will be developed across R10 that is more or less positive. The resulting voltage pulls the tuner oscillator back on frequency. L2 is an RF smoothing choke; C9 and C10 are filters.

RCA's All-IC AFT

The CA3064 all integrated circuit automatic fine tuning control circuit in a T0-5, 10-lead circular can, is in a separately shielded compartment at the rear of RCA's CTC46, 49 MAK video IF module. It contains (Fig. 5-20) a limiter amplifier, balanced detector, differential amplifier, AGC amplifier and voltage regulator. It is functionally similar to the previous CA 3044, but with input transistors and detectors instead of diodes.

The input (Fig. 5-21) is applied to the Q2 emitter follower stage through pin 6 and thence into Q3, which is in cascode with Q1, the feed (at pin 2) for the L1 primary winding of the discriminator transformer. The output of the transformer's tuned secondary supplies Q7 and Q13 with out-of-phase voltages through pins 1 and 3, respectively. A tertiary winding from L2 is connected to the base of Q1 through terminal 6 and to the bases of Q8 and Q12 that, together with Q7 and Q13, form a peak detector. The Q7 and Q13 emitter outputs pass to Q9 and Q11, which is, with Q10, a constant current driven differential amplifier with the collectors DC coupled to the tuner. Here, either Q9 or Q11 conducts more or less and delivers a corrective voltage that is used by the tuner for frequency

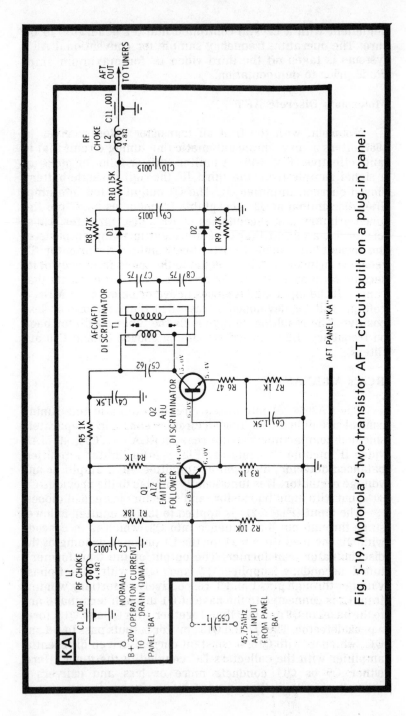

Fig. 5-19. Motorola's two-transistor AFT circuit built on a plug-in panel.

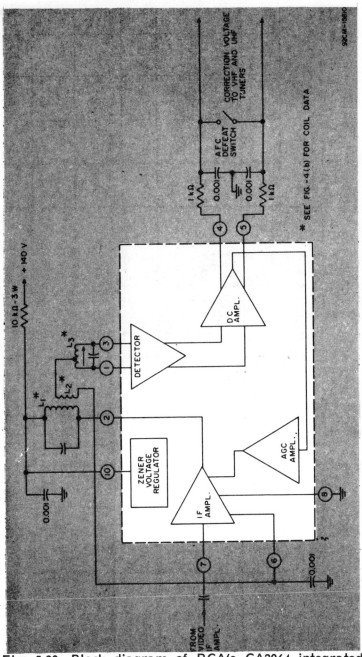

Fig. 5-20. Block diagram of RCA's CA3064 integrated circuit AFT.

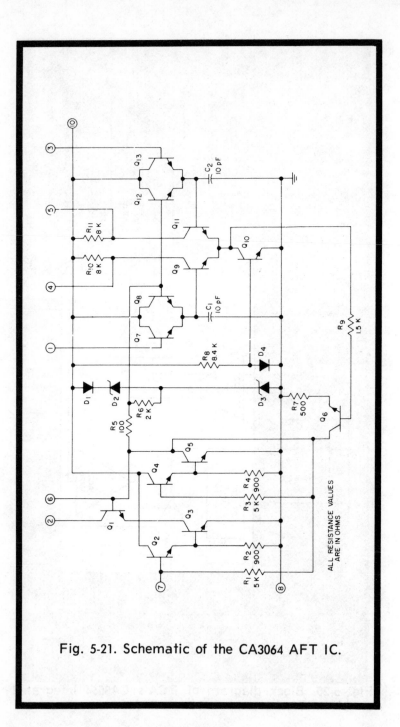

Fig. 5-21. Schematic of the CA3064 AFT IC.

correction. For instance, a negative frequency deviation causes Q9 to conduct more and a positive frequency deviation turns on Q11 harder.

AGC action comes from the collector of Q10 through R9 to the base of Q6. This transistor then helps regulate the bias for Q4 and Q7, thus limiting the operation of each transistor. Narrow band dynamic control is just over the range of plus and minus 30 kHz.

QUESTIONS

1. Draw a response curve showing the relative positions of the video IF and chroma subcarrier markers.

2. What are the two basic traps on **any** TV IF response curve and what are the frequencies?

3. Name the traps nearest the skirts of any IF response curve and identify as to frequency.

4. Which is at the higher frequency—the sound or video carrier, and by how much?

5. Most UHF tuners are tube-type tuners. True or False?

6. Some TV watchers say there is no real external tuner rejection of unwanted frequencies **below** 54 MHz. Is this true?

7. Name some favorable tuner characteristics.

8. What is the basic function of any tuner?

9. Cross modulation means what?

10. Do you think semiconductors make better RF amplifiers than tubes?

11. What's the difference between switch-type and turret tuners?

12. How do varactors work?

13. Describe the function of a color video IF strip.

14. Why doesn't the luminance bandpass on color receivers exceed about 3.5 MHz?

15. Write from memory (Table 5-2) the IF traps, carriers and sidebands.

16. Traps determine what characteristic of any swept response curve?

17. Name the three carriers that should appear on all TV response curves.

18. Does the lower adjacent-sound trap also trap out the sound carrier?

19. Why do you think series dropping resistors in vacuum tube receivers are inefficient?

20. Why do you suppose T3 in Fig. 5-15 is double tuned?

21. Who put the first video IF monolithic integrated circuit into a television receiver?

22. After reading the circuit description of TV's first IC video IF, do you think an IC will simplify or further involve TV servicing?

23. Do automatic fine-tuning control circuits differ as to 1) function, 2) circuits to make them operate?

Chapter 6

Video Amplifiers & Audio Systems

Video detectors and video (luminance) amplifiers are being given considerably more attention as chroma processing and picture tubes improve so that maximum video detail and resolution can be supplied to the overall system. Our initial block diagram (Fig. 6-1) is much like the one in Chapter 1.

DC SIGNAL COMPONENT

Norman Doyle and Don Smith of Fairchild Semiconductor say: "the DC component of a television signal contains information relating to the mean brightness of the scene being televised." So if there is no DC component, mean brightness is AC coupled and only picture content determines the average value of video voltage—and this isn't good. Therefore, you must either DC couple all video amplifiers to the picture tube following the video detector, or find some means of restoring the DC level.

In color receivers, there are **three** color difference voltages to contend with, and AC coupling could only produce a background tint proportional to the mean values of all three signals. In DC restoration or coupling, you fix a DC level and then position black and white levels with respect to it. And you must be sure, according to Doyle and Smith, that the contrast gain control and various signal conditions do not upset this fixed DC level. Also, an all-DC-coupled system will cause the

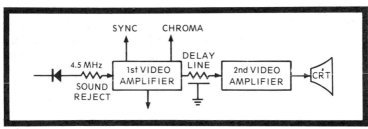

Fig. 6-1. Conventional video detector and video amplifier block diagram.

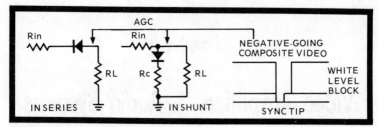

Fig. 6-2. Half-wave video detector diode in series and shunt, depending on bias, source and output impedances. Both pass only negative-going demodulated video as shown.

receiver cathode ray tube to go white to the point of blooming with no incoming signal because of negative carrier modulation, and the absence of carrier means "whiter than white." Black level clamping, therefore, is normally used in most systems, along with some slight capacitor coupling, at least in the luminance channels. With no incoming signal, then, the CRT is dark and excessive current isn't drawn.

To translate, getting undistorted television signals to the chroma and luminance amplifiers has always been a problem because of the characteristics of the video detector diode. And although the old faithful germanium diode in most systems works fairly well, it is basically nonlinear. Any diode detector uses unidirectional characteristics of doping and crystalline structure to produce an output voltage proportional to the level of detected modulation—less a 0.2-volt drop. The operation equation may be set forth in terms of current:

$$i_o = \{g/(1+gR1)\}\ e_i$$

where g is diode conductivity, R1 the terminating impedance, and ei, the input voltage. Diodes may be connected in series or shunt, depending on biasing and the source and load impedances (Fig. 6-2). Both diodes, in these examples pass only the negative half on the demodulated waveform. If an effective 2-diode arrangement, such as outlined in Fig. 6-3, were designed, there would be considerably less harmonic distortion because of full-wave rectification, and the carrier frequency would be doubled, cutting most distortion in half.

DIODE DETECTORS AND VACUUM TUBE VIDEO AMPLIFIERS

In tube receivers, only half wave diode detectors have been used for sound and video detectors as illustrated in Fig. 6-

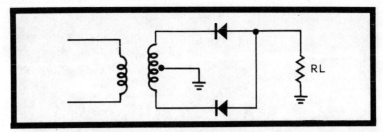

Fig. 6-3. Full-wave video detector diode circuit, which is not practical for color TV in the simple form.

4. Here, both diodes pick off positive or negative modulation for the respective circuits, and video passes over the 4.5-MHz carrier sound trap as described in Chapter 5. First video amplifier V304A is degeneratively biased with a 22-ohm cathode resistor. There is a series peaking coil in the plate circuit, along with additional peaking and LC resonant circuits in the coupling and grid circuits of the second and third video amplifiers. The plate and screen grids of the first video amp-

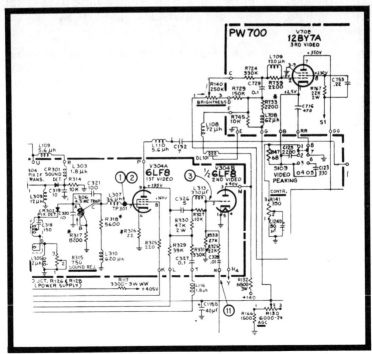

Fig. 6-4. Vacuum tube video amplifiers in the RCA CTC25 series.

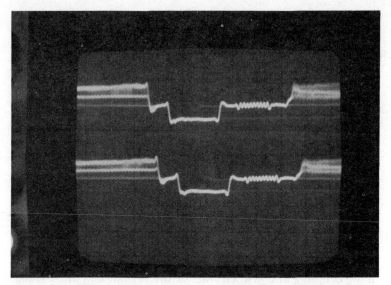

Fig. 6-5. If you want to check a delay line and see if it's actually working, here's how. The scope's time base is set at 2 microseconds per division. The delay is 0.8 microsecond; count it!

lifier supply signals to the AGC and sync circuits, plus chroma signals to the bandpass amplifier grid (not shown).

The second video amplifier grid circuit is rather heavily endowed with both noise cutoff and vertical interval blanking inputs, plus sync-stripped video from the first video amplifier. V304B passes the luminance signal through 0.8-microsecond delay line DL101 (Fig. 6-5), and into the grid of the AC-coupled third video amplifier, the driver for the CRT cathodes. Notice in the photograph that the trailing edge of the lower trace is delayed exactly the 0.8-microsecond interval, since the oscilloscope's time base was set for 2 microseconds per div. Squiggles (9 of them) following the horizontal sync pedestal are the 9 cycles of color sync burst. The scope's vertical amplifiers were set to alternating since this frequency is much too fast for a chopped input. Actually, we put the scope on 10 microseconds per division and used a X5 extender (now 2 microseconds per division) to precisely show these two signals across the delay line. Another very useful application for a good oscilloscope—and the only simple way to discover if a delay line is good or not.

High-frequency compensation is normally provided by **shunt** peaking with a coil, but **series** coil peaking isolates the effects of input and output capacitances, and so offers an advantage. Combined, shunt and series peaking permits higher gain than either used singly. Low-frequency compensation is usually added with an RC combination filter somewhere in the plate circuit of the video amplifier. Also, the value of the cathode (or emitter) bypass capacitor largely determines the low-frequency response of an amplifier.

BRIGHTNESS LIMITERS, BLANKERS & VIDEO AMPLIFIERS

In Zenith's C4030 video circuits (Fig. 6-6) composite video is coupled into second video Q205 principally through capacitors C213, C214, and partially through the contrast control. DC is applied through contrast R230 from the 24-volt line as forward bias, shown as 3.25 volts on the base of the transistor. Q205 has a 1.3K load resistor connected to the 24-volt supply line on one side of the collector, and series peaking coils L202, L204, and the L203 0.8-microsecond delay line in between. Signals out of Q205 go directly to the base of the Q206, the third video amplifier. The normal-setup switch in the emitter circuit removes video in the setup position.

Out of the emitter of Q205 is an RLC network that includes the brightness control and vertical blanker. The base circuit of Q205 contains the brightness limiter. This latter circuit senses excessive beam current by monitoring the current out the high-voltage tripler. The brightness limiter input is clamped by Zener CR214 and thus is protected against excessive high-voltage arcs. If the HV current flow rises above normal, Q204 acts as a partial or complete switch and drops the voltage from the 24-volt supply across 10K R228 so that the contrast control and, therefore, the base of Q205 receives less positive voltage or forward bias. The lower bias reduces the conduction of the second video amplifier and its signal current output and so controls the CRT beam current.

The vertical blanking circuit receives a positive pulse from a winding on the vertical output transformer. Emitter-follower Q207 applies the pulse to the emitter of second video amplifier Q205. At the end of each 15.2-millisecond vertical trace interval, the vertical output transformer furnishes an inductive kick of 23 volts p-p, turns on the Q207 vertical blanker, and the 18-volt pulse as its emitter firmly back-biases the emitter of the second video amplifier, thus cutting off the transistor. Across the 470-ohm resistor, also at the emitter, an 11.1-microsecond horizontal blanking pulse arrives following the 52.4-microsecond forward line scan. This pulse also turns off Q205 by back biasing.

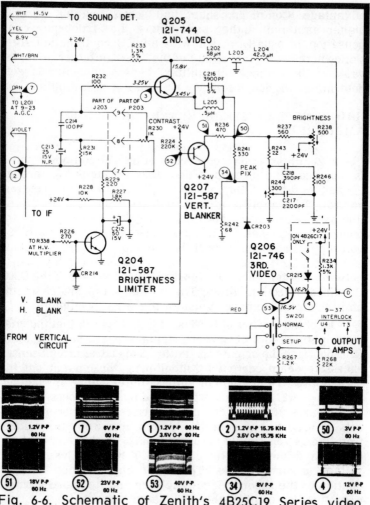

Fig. 6-6. Schematic of Zenith's 4B25C19 Series video amplifiers, as well as brightness limiter and unique vertical and horizontal blanking circuits.

Very simply, both horizontal and vertical blanking are achieved in a single stage with a minimum effect on the second video amplifier, since there is no overstressing caused by the simple cutoff bias. Video circuits must always be blanked at the horizontal interval, you may recall, because that is the time the horizontal sync pulse is transmitted as well as the 3.58-MHz color sync burst. The vertical interval of 1.4 milliseconds per field is used for reception of vertical, horizontal, and equalizing pulses.

Notice, also, the voltage dividers among the various transistors, the direct (DC) coupling, and the relatively few peaking and choke coils that are designed into the circuits. In dealing with transistorized receivers, you will find this is a normal condition rather than an unusual one. Once you understand transistor operation and have a good grasp of biasing and gain, such circuits should be considerably easier to troubleshoot. In our investigations, we have always found that in solid-state work a DC amplifier, triggered oscilloscope is indispensable. Sincerely, we doubt if you can get along without it. For instance, do you really have a meter that will measure 3.25 and 3.45 volts at the base-emitter of Q205? And if you can measure the voltage with a meter, is Q205 being blanked during the horizontal and vertical intervals at the proper times, and is its video output adequate? Worthwhile questions for, we hope, receptive minds.

As we have stated, germanium diodes used as video detectors have served useful purposes throughout the years, but are fundamentally nonlinear. And when current gain in any third or fourth-stage video IF amplifier reaches a certain point and begins to decline, the amplified output will be somewhat nonlinear. This is illustrated in Fig. 6-7 and is taken from Application Note AN-545, authored by Terry Kiteley, Motorola, Phoenix. Additionally, forward impedance characteristics of the detector diode's germanium material add to this nonlinearity.

These dual nonlinearities produce sum and difference beats called "tweets," which can always combine with the video signal or radiate back into the IFs and even to the receiving antenna, causing sync and video instability as well as external interference. TV manufacturers, Kiteley says, use various types of filters and shields to minimize these effects. But he also points out that a good low-level detection system without the traditional germanium diode would do away with these problems more easily.

In Fig. 6-7, the modulated video carrier is outlined and shaded at the bottom, while the output is strongly influenced by the curving dark line between diode and transistor h_{fe}, and is shown both as white level compression and sync pulse distortion.

In place of the diode, Kiteley proposes a doubly balanced, full-wave, synchronous detector (you'll see more of this when we're discussing chroma demodulators) wrapped up in a convenient 8-pin plastic in-line package with a conversion gain of 34 db, a video frequency response of 6 MHz, and video output of 7.7 volts p-p. The linear transfer characteristics are

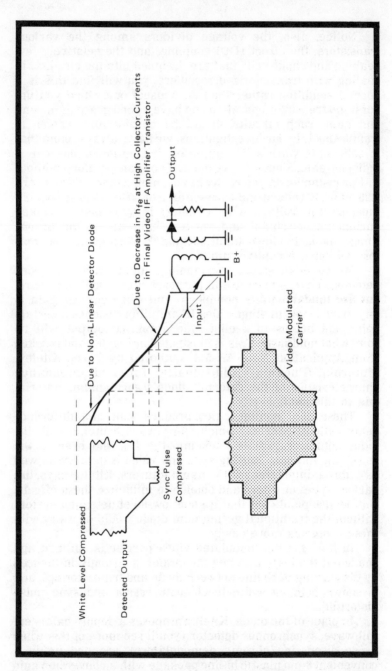

Fig. 6-7. Last video IF and diode detector transfer characteristics, showing the usual nonlinearities.

exhibited in Fig. 6-8. The line between transistor output and diode output is straight, with an undistorted output and neither sync nor video compression. A schematic of the unit, including IF input and video output, is illustrated in Fig. 6-9, while a block diagram for such a detector is exhibited in Fig. 6-10. The video modulation is amplified and limited, passed to a full-wave **multiplier** for delivery to a low-pass filter and the video output.

In the simplified schematic in Fig. 6-11, transistor Q7 is a forward biased constant-current source. Q1 and Q2 form a differential amplifier, with the base of Q2 at AC ground and the modulated carrier applied to Q1. Pairs Q3, Q5 and Q4, Q6 are the in-out phase clipped carrier-operated synchronous switches. Waveforms on the right side of the illustration show the Q1, Q3, and Q4 base inputs, and the resulting full-wave output developed across R1.

If an in-phase clipped carrier (Fig. 6-11B) is applied to Q3 and Q6, and the negative-going modulated carrier goes to the base of Q1, any possible Q1 conduction would back bias the Q3-Q5 pair of switches because of phase inversion, but turn on Q2 without phase inversion and Q6 would conduct. Current would, therefore, flow through R2 and produce an output. If Q1 were fully back biased, Q2 would conduct through differential action and turn on Q6 anyway. When the modulated carrier has larger positive incoming signals, Q6 is nonconducting because

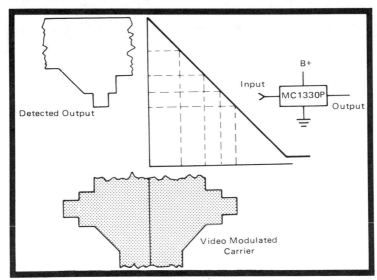

Fig. 6-8. MC1330P linear transfer characteristics. Notice there is no sync or video compression.

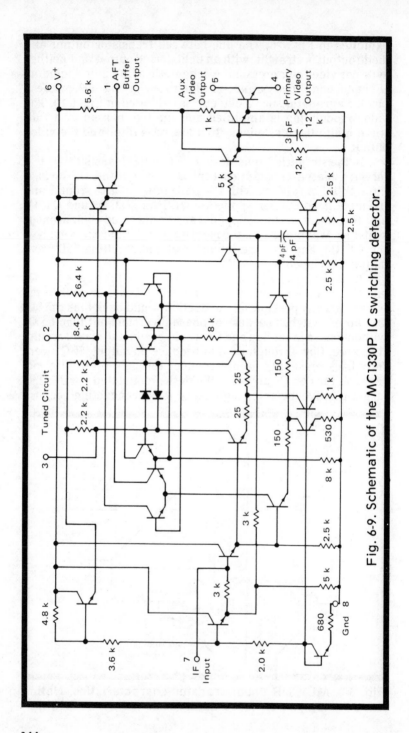

Fig. 6-9. Schematic of the MC1330P IC switching detector.

Q2 is off, while a negative voltage forward biases Q3 and produces an output across R1. With a change in clipped carrier phase, Q4 is turned on by Q2's conduction. Then, with a modulated carrier phase reversal, Q5 operates, since Q1 conducts negatively and forward biases its emitter, increasing the current across R2.

This is called synchronous detection. First one and then the other switches operate as clipped carriers while modulated carriers go through a phase reversal and develop currents across R1 and R2. This action, of course, causes the detector to be switched at twice the carrier rate, leaving no original carrier present, but simply modulated pulses that are double those of the original carrier frequency.

It is important that switching voltages contain only carrier energy and not color information. Therefore, this external circuit must have a filter such as L3 and C10 in Fig. 6-12. The figure shows two integrated circuits, the first two video amplifiers (MC1350), including AGC amplifiers and bias supplies. The second MC1330 device (schematic in Fig. 6-9) is the circuit we've been looking at, which also contains the third video IF. As you can see in Fig. 6-12, there are still coils to be tuned (the T1 coupling transformer between integrated circuits), so we don't have a transformerless video IF yet. At the outputs of MC1330 are two out-of-phase video signals, either of which may be used. The two signals are simply the emitter and collector outputs of the output transistor shown in Fig. 6-9.

FM DETECTION & AUDIO AMPLIFIERS

TV receiver audio systems have been taken pretty much for granted these last 20 years because there really wasn't much to them except rather inferior sound delivery. It all began with the **good** Foster-Seely discriminator that was somewhat susceptible to noise. Sound systems evolved to the popular ratio detector which was more noise immune, then

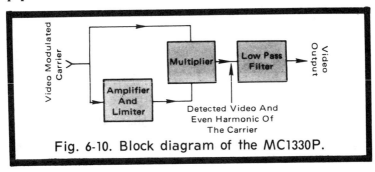

Fig. 6-10. Block diagram of the MC1330P.

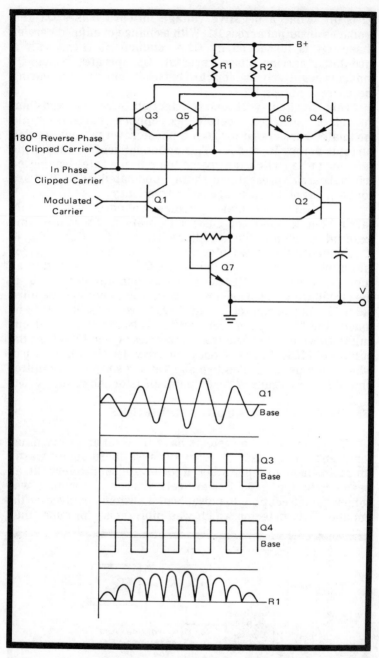

Fig. 6-11. Simplified schematic of the MC1330P and switching waveforms.

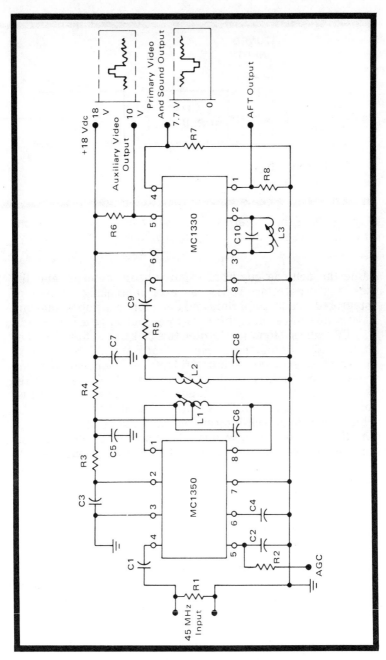

Fig. 6-12. IF system using an MC1350 (two IF amplifiers) and an MC1330 (final IF and synchronous detector).

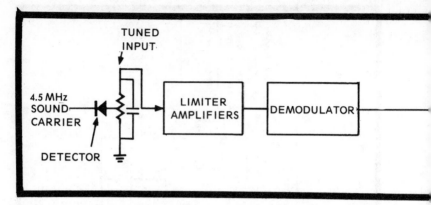

Fig. 6-13. Block diagram of a typical audio system showing detector diode, amplifiers, demodulator, and two

came the pentode quadrature gated-beam detector and the locked-grid detector. Today, an ever increasing number of integrated circuit amplifiers and detectors are appearing, and at least one all-ceramic throw-away output stage is in use.

TV audio systems are limited to a 50-kHz bandpass, rather than the 150-kHz bandpass reserved for FM and FM stereo. So you can't expect overwhelming fidelity nor sound-surrounding output. Nonetheless, with the advent of ICs and better designed push-pull output stages that no longer use transformers, the quality of TV sound has improved remarkably. This is due to the self-limiting characteristics of the internal amplifiers, more precise demodulation, better de-emphasis and tone networks, more powerful and efficient outputs with less built-in distortion, and good impedance matches with an output coupling capacitor in place of the usual lossy transformer. The large bandpass of more recent power transistors has aided, too.

A block diagram of a conventional TV sound system is shown in Fig. 6-13. Sound is picked off at the video detector by a sound detector diode, put through a 4.5-MHz FM audio carrier tuned circuit and fed into one or more limiting amplifiers. After FM is demodulated and de-emphasized to counter transmitter pre-emphasis, the audio is amplified in two stages and coupled to the speaker. The "cheapies" chop this sound procedure down somewhat, but all such subsystems are basically the same. The big difference exists in FM-to-audio detectors, and the less expensive receivers almost inevitably use a type of gated-beam detector, since demodulation requires a single tube or perhaps one or two transistors.

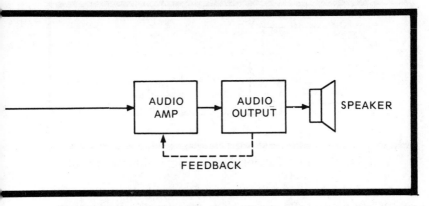

stages of audio amplification. The speaker in more recent receivers may be capacitor or transformer coupled.

Otherwise, the block shown is just about what you should expect to find in any receiver on the market.

Discriminator

This forerunner of FM detectors dates back to much earlier days when frequency-modulated radio was a novelty, so obviously it isn't especially recent or unique. But the sound it produced was good, and some **radios** today may still be using the idea, if not the circuit.

Referring to Fig. 6-14, both primary and secondary circuits are LC tuned to the 4.5-MHz incoming carrier, and the signal is electromagnetically and capacitively coupled from primary to secondary. If the signal is exactly on frequency (carrier frequency), each of the secondary diodes receives 90 degree out-of-phase signals, and the rectified output currents develop equal and opposite (canceling) voltages across R2 and R3. When the frequency of the input signal varies below or above 4.5 MHz, the diodes—180 degrees out of phase with one another—develop relatively positive or negative outputs that reflect the degree of higher or lower incoming frequencies (modulation). The changing carrier frequency generates audio voltages that are proportional in amplitude to the FM broadcast deviations. You can always recognize an F-S discriminator circuit by the two detector diodes in parallel and the small capacitors.

Ratio Detector

The unbalanced ratio detector (Fig. 6-15) looks much like the F-S discriminator, except it has series instead of parallel

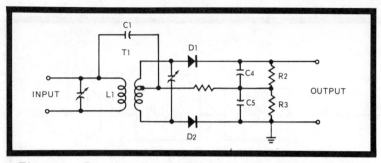

Fig. 6-14. Durable Foster-Seely discriminator circuit.

diodes, and includes a big capacitor across load resistors R1 and R2 to ground. Because of C4, the circuit's load voltage cannot follow small changes in the input signal and is, therefore, not sensitive to incoming amplitude variations. Phase differences between secondary and primary, and on either side of the tapped transformer secondary, are the same as in the F-S discriminator. However, since the detector diodes are in series, they generate a constant sum total of output load voltages at all times, but their ratio varies with the input signal deviation. The output is a changing ratio across the output resistors rather than a pure voltage difference. This is why such a circuit is called a **ratio detector**. Since the circuit always has a DC output, coupling capacitor C3 is installed to supply an AC-only audio voltage to the de-emphasis network and the audio gain potentiometer.

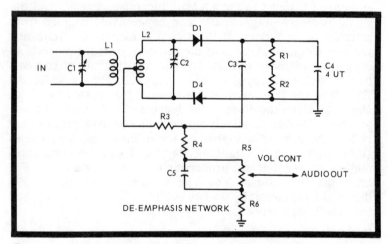

Fig. 6-15. An F-S successor, the unbalanced ratio detector which is less susceptible to AM noise.

In both the F-S discriminator and the ratio detector, as the input frequency passes above or below the resonant frequency, current in the transformer secondary lags or leads the induced voltage. And since voltages across the secondary coils lag the secondary current by 90 degrees, the voltages across these coils are then 90 degrees out of phase with the primary voltage when the received frequency matches the tuned resonant secondary.

Vacuum Tube Sound Subsystems

RCA's CTC25X chassis sound system is most representative of the middle '60s era (Fig. 6-16). Sound is picked off by a positive demodulating detector diode in the plate circuit of the third IF and brought to the grid of the sound IF (V201) through an RLC network following the diode and another blocking capacitor, plus additional series and shunt coils that reject video and sync and form a tuned 4.5-MHz circuit.

The 6EW6 tube is a sharp cutoff pentode used in gain controlled audio and video IF stages. It features controlled plate-current cutoff and a high transconductance of 1400 micromhos. It's relatively large plate swing, along with the other characteristics, provides considerable gain so that only one stage is needed before demodulation. The sound IF (T202) transformer is double tuned for broad bandwidth, with a resonant circuit in its secondary, which is a good impedance match for the grid of the following sound demodulator.

The 6HZ6 demodulator is a sharp cutoff pentode with two independent control grids. It may be used either as a Class A amplifier or as an FM sound detector in FM and television receivers in either parallel or series-connected heater strings. Used by both Zenith and RCA in much the same configuration, the circuit is similar in many respects to the old 6BN6 gated-beam detector, which had a rectangular cathode surrounded on three of its four sides by a focusing anode. Electrons leaving the cathode form a sheet-beam for the focusing anode, go through a narrow slot in the accelerator to the first control grid. If the control grid is slightly negative or positive, electrons may pass, but otherwise they are blocked and return to the cathode. The control grid, meanwhile, is saturated by large positive signals or cut off during negative signals and so limits the received FM signal amplitude. When the control grid allows electrons to pass, they go to a second accelerator, the screen grid, and on to the quadrature grid, connected to a 4.5-MHz high Q resonant circuit consisting of the RLC network between ground and pin 7.

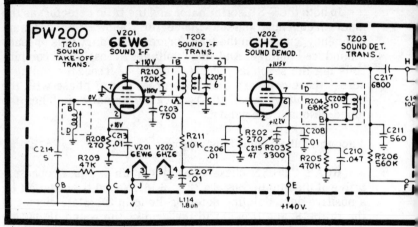

When the incoming signal is exactly 4.5 MHz, the tuned quadrature grid lags this frequency by 90 degrees. Higher frequencies cause larger phase lags and lower frequencies lesser lags. The combined effect of the two grids cause different pulse current widths to arrive at the detector plate, making the plate current a linear function of FM deviation. Audio voltage is developed across load resistor R206 and passed through a de-emphasis RC network and the volume control to the grid of the audio output. Notice that the tone control is in parallel with half of the volume control and is simply an RC series pair of components that can attenuate the higher frequencies, with possibly a boost for the low ones.

The 6AQ5 audio output cathode is heavily bypassed, thereby allowing the tube to work virtually at full potential and deliver several watts of output. The suppressor grid is at cathode potential, while the screen is supplied by the +140-volt

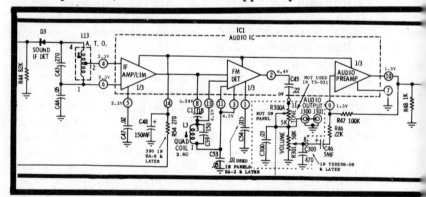

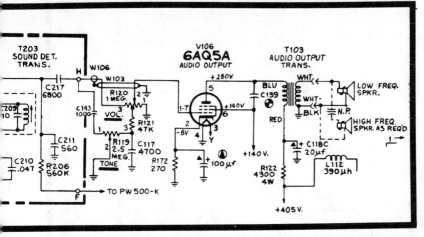

Fig. 6-16. RCA CTC25X vacuum tube sound subsystem circuits.

source. Capacitor C139, between plate and screen, is a basic troublemaker if its transient suppression and stabilizing function is disturbed, especially by a short. The primary of output transformer T103 often opens because of the 405-volt supply and high current demands of the output tube.

Motorola's CTV7-8 audio subsystem (Fig. 6-17) uses an IC and a push-pull capacitor-coupled output. However, the solid-state FM detector **does** have a quadrature coil like the 6HZ6 tube, and so also is known as a quadrature detector.

The FM sound takeoff point is in the collector of the third video IF transistor. The 4.5-MHz signal passes through the sound IF detector into the (L13, C43, 44) 4.5-MHz tuned input

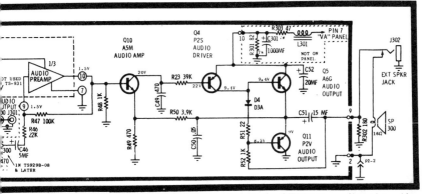

Fig. 6-17. Schematic of Motorola's CTV7-8 audio circuits.

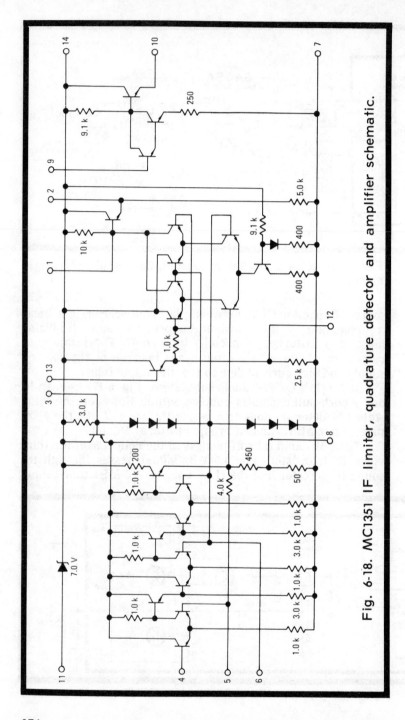

Fig. 6-18. MC1351 IF limiter, quadrature detector and amplifier schematic.

circuit. The IC input is terminal 4, since terminal 6 is actually a bypassed bias supply.

The IC itself is also quite interesting and is a Motorola type MC1351. It is a dual in-line monolithic IC with 22 transistors, 7 diodes, a zener diode, and 21 diffused resistors. It has three limiting differential amplifiers (Fig. 6-18), voltage regulation by the 7-volt zener and diodes D1-D6 plus Q10, an effective means of quadrature demodulation, current-multiplying drivers and an emitter-follower final stage serving as a low-impedance preamplifier.

The three differential amplifiers, each with a gain of 10, are buffered by emitter-follower outputs. The amplifiers are limited in output swing for excellent AM rejection and furnish a total 60 db gain over the frequency range of 1 to 12 MHz. Following the six temperature-compensating and regulating diodes are another two pairs of cross-coupled differential amplifiers supplied by a single transistor current source and another feeder pair, again differential in configuration. The right-hand transistor of this pair has a regulated DC voltage on its base, while the basic FM signal is delivered to the base of its sister transistor on the left from the third buffer transistor following the three limiters.

On the CTV8 schematic (Fig. 6-17), a quadrature coil and shunting 150-pf capacitor form a tank circuit, the Q of which determines the peak separation of the detected output, noise characteristics, and output voltage swing of the detector. Since this is a quadrature system, the tuned circuit will lag the incoming carrier by precisely 90 degrees. The output, through the emitter follower transistor to pin 2, is twice the input through pin 13, the initial emitter-follower and the 1K limiting resistor. The time constant of the 10K resistor in the right differential amplifier and the 0.015-mfd capacitor (normally the de-emphasis network) shunts the twice multiplied incoming frequency and any noise pulses to ground. Higher frequencies result in a greater phase lag and lower frequencies produce a smaller phase lag, with maximum output at 0 degrees and-or 180 degrees.

These differential amplifiers are, in reality, switching circuits that combine the varied pulse transitions through the emitter follower transistor (pin 2) and develop an audio output that is a linear function of frequency deviation. The recovered sound is then passed through pin 9 into the audio preamplifier current driver transistors and out through the emitter follower at pin 10, with some stabilizing feedback returning to pin 9. Further amplification is provided by Q10 and Q4. Q4 is a class

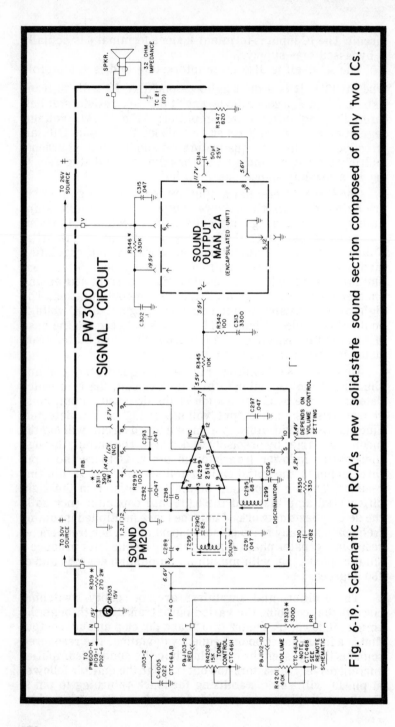

Fig. 6-19. Schematic of RCA's new solid-state sound section composed of only two ICs.

B (approximately) push-pull output coupled through a 150-mfd capacitor directly to the speaker.

In the complementary symmetry output stage, as it's called, the NPN-PNP coupled emitters are quiescently biased at 10.7 volts. The upper NPN base has 11.2 volts and the lower PNP base has 10.2 volts. Since the NPN's base-to-emitter voltage difference is +0.5 volt, and the PNP's base-to-emitter voltage difference is -0.5 volt, both transistors are forward biased to within 0.2 volt of conduction, also an antidote for crossover distortion. A positive audio voltage will drive Q5 into conduction, and a negative audio voltage will pass DC forward-biased D4 and drive Q11 into conduction. The outputs are then coupled through C51 to the speaker. Resistor R50 supplies feedback to the emitter of Q10 for DC stability and to minimize distortion at high output levels. R51 and R52 are simply the voltage dividing network for the base of Q11, while R302 is part of the load and charging path for C51.

RCA's 2-stage IC thick film audio processor is represented in Figs. 6-19 through 6-22. Since an entire integrated circuit is obviously more expensive than several individual transistors, manufacturers must have reasons for going substantially all-IC. The CTC46 and CTC54, two new RCA solid-state chassis, have supplied many of the answers, except cost, and we'll probably never know this. However, the circuit on board PW300 consists of only three basic entities: one monolithic IC299, one phenolic encapsulated thick film IC, and one speaker. If you count the peripheral components, they amount to about 22 passive pieces—a far cry from almost hundreds used in other sets. So the initial answer here is economy of parts and low labor costs, since the PM200 sound section is a plug-in module, and so is the MAN 2A.

IC299 is a CA3065 complex multifunction IC (Fig. 6-20) with an IF amplifier-limiter supplying an FM detector. The detector output goes to an electronic attenuator, to a buffer, and then to an audio driver and the MAN module. The block diagram, of course, is a "typical" circuit application that makes the routine look easy. But let's go to the schematic (Fig. 6-21) of this little 14-pin IC and you'll get a different perspective.

Incoming sound, applied to terminal 2, passes through the Q11 emitter to the collector of Q12 without inversion, is coupled emitter-to-base to another set of differential amplifiers, Q14-Q15, then delivered to a third set consisting of Q18 and Q19 still in phase. Reference zener D2 and regulators Q1, Q3 and Q4 supply collector regulation for Q12, Q15 and Q19, and detectors Q22 through Q27. The bases of Q17 and Q20, are clamped by D7,

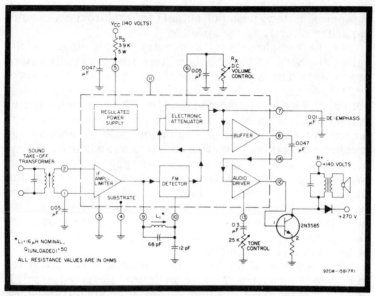

Fig. 6-20. A block diagram of the CA3065 in a typical circuit application.

a limiter and current source, respectively. The Q18-Q19 differential pair again amplifies without phase change through Q21 (with feedback) and to the Q22 and Q27 buffers through the L1-68-pf tunable discriminator circuit (Fig. 6-19) Q23 and Q26 are peak detectors that supply differential amplifiers Q24 and Q25 with detected FM that is compared against the resonance set up by the discriminator circuit so that output currents are proportional to the input FM deviations. After amplification by Q7-Q8, signals are applied to the electronic attenuator.

Terminal 6 is the DC volume control connection, while bias for Q5-Q6 and, subsequently, bias for the bases of Q8-Q9 and Q7-Q10 is supplied from the emitter of Q36 out of the regulated chip supply. Q29 is the emitter constant-current source for Q9-Q10. If you look at these two pairs of differential amplifiers, you will find that, except for the emitters, they are in parallel. The only mobile connection is to the emitter of Q7-Q8 and this is clamped by the limited voltage from the power supply, base bias and the constant current feed from Q29. The output through Q2 and terminal 8, therefore, is electronically and automatically attenuated regardless of incoming signals.

Low-impedance audio is delivered from terminal 8 to terminal 14, the input to buffer Q33 and audio drivers Q34 and Q35. Transistors Q30 through Q32 are simply another regulator, this time for Q33. Any external changes are sensed

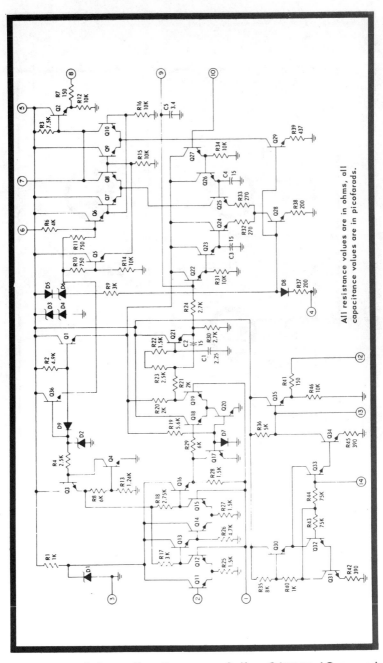

Fig. 6-21. Schematic diagram of the CA3065 IC sound detector.

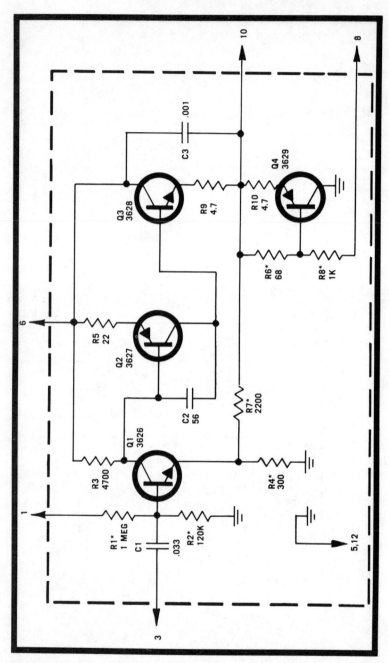

Fig. 6-22. Schematic of the MAN 2 Module phenolic version.

182

by the base of Q30 which increases or decreases the emitter current and also the collector current of Q32, which finally reaches the base of Q31 where the change is resistively controlled by the conduction of this transistor.

The MAN 2A module (Fig. 6-22) receives its input from terminal 12 of PM200 through a 0.033-mfd capacitor between base bias resistors R1 and R2. The signal is amplified through Q2 with a little degenerative feedback to the base of Q2. Transistors Q3 and Q4 constitute a Class B push-pull amplifier, with Q3 taking incoming voltages from Q2; Q4 requires negative polarity inputs on its base or positive inputs on its emitter. Feedback applied is through terminal 8, following the 50-mfd capacitor at output terminal 10.

QUESTIONS

1. What would the face of the CRT look like in an all DC-coupled system with no incoming signal?

2. Why is black level clamping preferred?

3. Are video detector diodes quite linear?

4. Why is there a luminance delay line, and how much is the delay introduced?

5. What is the only way a serviceman has to check a luminance delay line accurately?

6. In the newer receivers, where the vertical and horizontal blanking pulses applied?

7. What happens in the luminance amplifier during vertical and horizontal blanking times?

8. What are "tweets"?

9. How can we be rid of tweets?

10. What's the bandpass of most TV audio sections as opposed to those of FM stereo?

11. What was the first good FM audio detector?

12. Do IC detectors and amplifiers produce better sound than discretes?

13. The ratio detector diodes are connected in p_____, while the discriminator's diodes are in s_____?

14. What is the basic difference between the ratio detector and discriminator?

15. What is the fundamental principal of the gated-beam detector?

16. What's different about Motorola's output stage in Fig. 6-17?

17. Can you guess how Q4 in Fig. 6-22 works?

Chapter 7

Sync & AGC Circuits

All television receivers have sync separators, most have AGC, but many do not have noise gates. The principle of the sync separator (Fig. 7-1) is simply to strip the video from the composite waveform and amplify sync pulses above the black level. Out of this action is derived the serrated vertical integrated (slowly and successively charging capacitive) pulses that time the vertical oscillator when it is not freewheeling or "flywheeling." These charging pulses come from the transmitted six vertical pulses that are surrounded by horizontal and equalizing pulses, but the relatively long duration of the vertical pulses allows the integrator to charge in steps, while filtering the shorter pulses to AC ground. The horizontal pulses, of course, appear during each line scan at the 52.4-microsecond rate, while the repetition rate of the vertical pulses transmitted during field times is 31,500 per second or inversely in time at about 32 microseconds. All horizontal, equalizing, and vertical pulses are completed in a single field of 1.4 milliseconds, half the frame rate of 30 complete pictures per second.

Horizontal pulses are merely RC differentiated to potentially sharpen the leading edge and filter out vertical frequencies. They become rather sharp rectangular pulses and provide precision timing for the horizontal oscillator. You can well imagine what happens when vertical pulses find their way into the horizontal sweep circuits, or horizontal pulses into the vertical sweep. Here is a pointed instance where a recurrent sweep oscilloscope is utterly useless, since there is no calibrated time base to determine the interfering frequency.

There's really not much more to a sync separator. Usually, it's a single-stage tube or transistor circuit that is biased sufficiently to clip the video, and the dual outputs (vertical and horizontal sync) are sent on the way. When troubleshooting vertical or horizontal circuits, it is a good idea (if you have a dual-trace oscilloscope) to hang one probe on the output of the sync separator and work through the defective sync circuit with the second vertical input. This way you can tell if there's any frequency deviation and always have a ready reference in case something peculiar turns up, such as the vertical or horizontal starting up all by itself. It's happened!

AGC is something else. The whole principle began with simple diode rectification of video which produced a DC feedback to the IF and RF amplifiers. Naturally, this has changed radically since the 1950s, and there are a number of systems in use today, most of them what is called **keyed AGC**. Very simply put, a keyed system operates only during the horizontal retrace time and then only on the tips of the horizontal sync pulses. Although every keyed AGC amplifier is lightly biased by some DC voltage, pulses from the flyback transformer are usually rectified by a diode to supply the operating voltage for the keyer. As horizontal sync pulses are received, the flyback transformer keys the AGC stage on, resulting in an output. The larger the amplitude of the incoming sync pulses, the greater the keyed AGC output, and the more controlling effect it has on the receiver's IF and RF amplifiers. In this way, tuner and video IF stage conduction is easily managed and the system offers not only marked resistance to airplane flutter but also gives some protection against noise pulses. Possibly in the future, AGC systems will combine keyed operation and a system based on **an average AC video**. The two would complement one another where each is individually least effective.

Noise gates are not always included in the less expensive receivers, and sometimes you're inclined to wonder if they really work or if they're just there for nuisance value. However, were it not for noise circuits, there'd be considerably more interference appearing on color television screens than there already is. Unfortunately, noise circuits can produce some rather tough troubles that seem to defy rapid detection.

Originally, noise stages (or circuits) were put in receivers to black out sync and perhaps video (this is what we're driving at when we show a noise gate going to the AGC in Fig. 7-1) when overriding noise pulses were present. Probably many people think this is still completely true, but it isn't. Just like the rest of today's more advanced television receivers, noise circuits are now definitely associated with AGC, and hardly with sync at all. In one of the newest sets, noise controls the AGC shift from one strength signal input to another so that RF amplifier conduction is increased or decreased for minimum noise and to prevent mixer crosstalk (RCA, if you're curious). So, in this instance, there's little talk about sync, but much

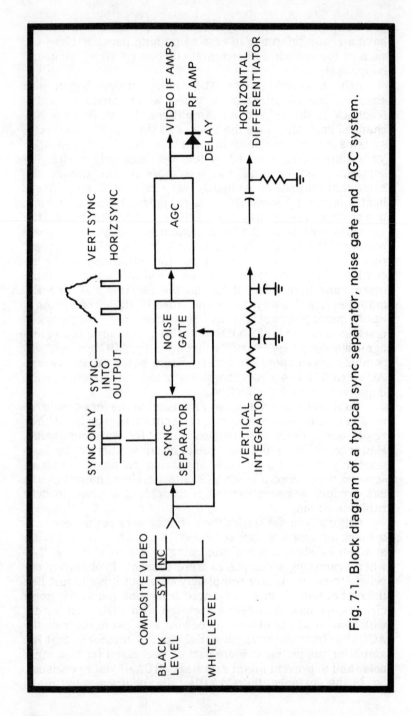

Fig. 7-1. Block diagram of a typical sync separator, noise gate and AGC system.

more interest in detecting increases in AGC action that affect a receiver front end and cause viewing problems that can be prevented by selected settings of the noise circuit.

By the way, although we're told to look at a noisy station and set noise controls for the least color snow, we have found that an oscilloscope reading the composite video signals before chroma takeoff can also show you where and when to stop tuning the noise potentiometer. It's at the point where you have a full amplitude video waveform and no sync compression. The author, by the way, uses the composite video signal on a strong station to set the AGC in all receivers if they have the usual DC control. Some CRT sync bends are hard to see. The oscilloscope is precise.

TUBE-TYPE AGC, NOISE, & SYNC SYSTEMS

At first glance, these are puzzling circuits (Fig. 7-2), but basics still apply. An LC series circuit resonant at 376 kHz from the grid of the first video amplifier supplies positive sync pulses at a horizontal rate to the grid (pin 6) of the keyed AGC

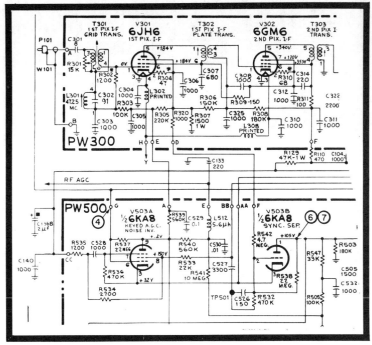

Fig. 7-2. Schematic of the vacuum tube keyed AGC, noise and sync circuits used in RCA receivers.

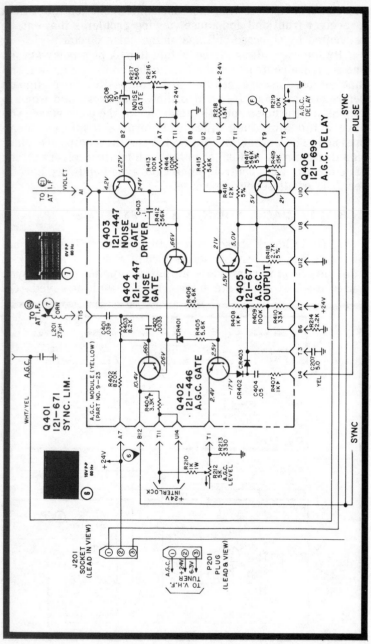

Fig. 7-3. Schematic of Zenith's 4B25C19 sync-AGC-noise module.

and noise inverter. Sync information included in the full video signal arrives at the screen grid (pin 8) through R533 and continues on to the sync separator through C526 and R532. Over regular variations in the amplitude of incoming video, the screen and control grids ride together and produce more negative voltage at the plate of V503A for larger incoming video.

Keying from the flyback (with filtering) is applied to the plate circuit of V503A, where the tube is also biased from the resistive string on PW300 through terminal E. If a large spike of noise appears, the grid and suppressor work hard, the positive-going noise spike is inverted and this drives the AGC amplifier into nonconduction, cutting off the picture and possibly also the sync, although the screen grid of the keyed AGC tube would have to be driven heavily for that. It probably depends on the size of the disturbing spike. The AGC set control determines the bias of both cathodes of the keyed AGC tube and sync separator.

The AGC IF negative control voltage finds it way to V301 through R302 and the grid circuit, again going into the board through terminal E. The RF AGC delay is accomplished by R540 and C530, with a charge delay of 5.6 milliseconds, and the usual drop through a 560K resistor.

The sync separator bias comes through terminal D on PW300. The bias is probably slightly affected by the AGC action of V503A. Plate voltage is only 73 volts with respect to cathode, so there's not too much swing. Also, there's a 22-meg resistor between plate and grid for added bias control. The capacitors and resistors aid the grid-plate action of V503B, which strips the horizontal and vertical sync information from the overall video signal. Relatively pure horizontal and vertical sync pulses appear at the plate of the separator. The grid is fed through terminal BB, C527, C526, and R532. Bias on the grid of V503B also should be fairly high since R538 and R542 act as a grid bias voltage divider; R538 is connected to a potential of 105 volts. We would suspect bias to measure around 30 volts, since the grid normally has to be slightly more negative than the cathode for satisfactory tube operation. V503B produces a single output and separate vertical and horizontal sync pulses are derived by low-pass (integrator) and high-pass (differentiator) filters.

SOLID-STATE SYNC, NOISE, & AGC SYSTEMS

In the Zenith Duramodule (Fig. 7-3), **negative** horizontal sync pulse spikes are applied to the module at A1 and the emitter of the noise gate driver, through R406 to the base of

AGC gate Q402, driving this transistor into conduction which is proportional to the amplitude of the sync pulses. Pulses from the flyback transformer key CR402 on, supplying the collector operating potential for the AGC gate. CR403 rectifies the positive portion of the gate output and C207 filters out any AC so that the base of Q405 now sees a more positive DC in addition to the forward bias already applied from the +24-volt line.

The AGC output stage has a certain amount of applied collector voltage also from the 24-volt line, which also supplies Q405's emitter with constant voltage and, at the same time, furnishes operating voltage for both the emitter and base of Q406 through R416 and divider R417 and R419. The AGC delay control completes the current path to ground and firmly divides the voltage drop at the base of Q406. In turn, the AGC level set controls the voltage and, consequently, the conduction point for the emitter of Q402, the AGC gate.

With AGC gate Q402 operating, a positive voltage turns on AGC output Q405 and its emitter rises above 5 volts. When this occurs, a positive AGC voltage is transferred to the IFs for forward, current-limiting bias. At 6.7 volts, transistor Q406 now begins to conduct, affecting the RF amplifier after the more than one volt delay, since the Q406 emitter must be 0.7 volt more positive than the base for Q406 to conduct.

The sync separator, meanwhile, takes positive video from terminal T15 through capacitors C401 and C402. Normally conducting diode CR401 is back biased by each negative incoming horizontal sync pulse tip. Video coming through C401 sets up a relatively negative voltage on the base of Q401 so that only the tops of the incoming positive sync tips overcome the bias and cause the sync limiter to conduct and pass sync information to the collector. During times of no signal input, CR401 supplies part of Q401's emitter bias.

Noise gate and driver have a bias set control that obviously can affect both sync and AGC since the conduction point of Q402 will consume negative sync pulses, and the action of the Q404 noise gate will upset the sync limiter bias and produce distortion. In such circuits, beware of hasty adjustments.

Noise gate control R216 is set so that the noise gate driver is back biased and will not conduct on normal signals. However, if a large negative transient appears at the emitter of Q403, the transistor will be forward biased and conduct, sending a signal through C403 and R412. This voltage is **not** inverted and so turns **off** the noise gate itself. Transistor Q404 then stops conducting, opening the emitter of Q401 so that it cuts off the sync limiter for the duration of the transient. Therefore, little or no AGC and no sync are generated during the nanosecond or microsecond interruption. In other words, you're missing only part of a line anyway, and the eye is normally too slow to follow any such transistory occurrence.

SYLVANIA'S SPECIAL AGC CIRCUIT

In the D12 Gibralta chassis (Fig. 7-4), collector supply voltages for the AGC gate are rectified pulses from the flyback transformer. Negative-going video from the first video amplifier drives Q304 into conduction during reception of horizontal sync pulse tips. The point of conduction depends on the setting of AGC control R352 at the bottom of the diagram. This control is bypassed by C308 to avoid amplifier degeneration.

With no incoming video, the base of Q304 rises to 2.75 volts to limit AGC action. As sync tips arrive in varying amplitudes,

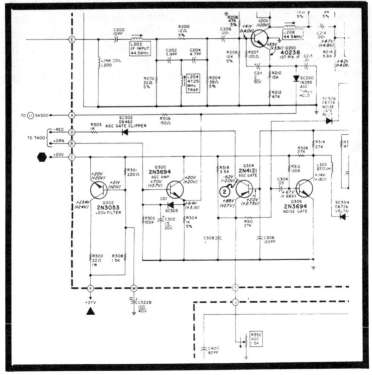

Fig. 7-4. Sylvania's special AGC, IF and RF control circuits. (Courtesy Joe Thomas)

Q304 conducts more or less and delivers pulses of proportional amplitudes back through SC302, across the D-B windings of the flyback transformer, and then to the base of AGC amplifier Q300. The large R302-C302 RC filter supplies a time constant of 1.5 seconds, sufficient to remove AC and permit only a varying positive DC drive for the unity gain emitter follower AGC amplifier.

At maximum gain, Q304 develops no positive bias, so the AGC amplifier does not conduct. But first IF stage Q200 is now constant at about 4 volts, while the emitter of the IF draws about 3 milliamperes through AGC threshold diode SC200 and AGC amplifier emitter resistor R304. This current sets a level of some 3 volts at the emitter of AGC amplifier Q300 and establishes a bias for the base of the AGC-controlled RF amplifier in the tuner.

With increasing signal input, the AGC gate develops a positive voltage and draws more emitter current through R304. This, in turn, reduces the current through IF amplifier Q200, and lessens the gain in the reverse AGC mode. At the same time, the voltage across R304 is constant; therefore, biasing the RF amplifier into maximum gain. With further increases of AGC voltage, current through the AGC amplifier increases and current ceases through Q200, so diode SC200 opens. Series resistors R210 and R212 become active and provide current control for the Q200 emitter with SC200 open and limit the maximum AGC gain reduction in the IF range, specifically determined for the best noise versus mixer overload. Further AGC increases will develop additional voltage across R304, delivering more forward bias to the RF amplifier.

SEMICONDUCTOR AGC DESIGN CONSIDERATIONS

The following discussion is based on material originated by Karol Siwko in Batavia, in which he analyzes IF response problems in semiconductor receivers and especially those coincident with automatic gain control.

Pole shifting (changing the IF response with gain) has been the usual design goal in vacuum tube receivers, where changing the input impedance altered the gm in a pentode. When transistors arrived, IF amplitude, phase response, stability, and gain had to be further dealt with, plus predictable AGC performance. Bipolar small-signal transistor gain can be controlled through output and input resistance by forward AGC, regenerative feedback, and also external damping diodes for frequency and circuit Q response.

Curves showing strong signal and weak signal responses are drawn in Fig. 7-5A for video IFs and at the intercarrier sound detector, Fig. 7-5B. The solid curve in each case demonstrates conventional response for strong signal reception, while the dashed curve is the signal response modified for weak signal reception.

When faint signals are being received, the picture carrier is shifted to the peak of the response curve for the following advantages:

1. With the signal rise time reduced, the overall noise is down and, with most picture content in the lower frequencies, the bandwidth reduction attenuates the high-frequency noise. Also, only low energy in the picture is reduced, and this produces an improved signal-to-noise ratio.

2. Maximum gain appears at the best fine tuning point, and this extends the range of the automatic frequency control, while all traps maintain their design functions.

3. The chroma carrier is moved further down on the response curve relative to the video carrier, producing less chroma output through the color level circuits, which also attenuates color noise during weak signals.

Also on feeble input signals, the sound intercarrier level is raised proportionally with that of the video carrier to maintain the sound limiting sensitivity, plus keep the center frequency amplitude response down so that the video sideband and intermodulation noise in the sound channel can be reduced.

A good quality TV IF amplifier should use at least five poles in the bandpass for sufficient compromise among

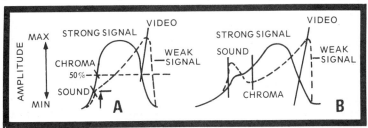

Fig. 7-5. The sound and video IF response characteristics rise ideally and proportionately with a shift from strong to weak incoming signals. Video rises in proportion to sound on a weak input at the video IF curve (A). Sound IF detector response (B) showing the sound and video rising and peaking on very weak signals. (Courtesy Karol Siwko)

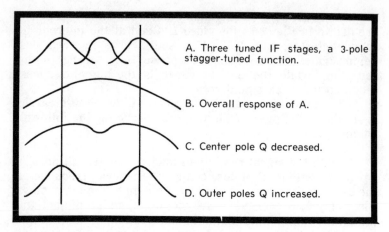

A. Three tuned IF stages, a 3-pole stagger-tuned function.

B. Overall response of A.

C. Center pole Q decreased.

D. Outer poles Q increased.

Fig. 7-6. Three poles in stagger-tuned stages, and what happens when they are combined or the Q is reduced or increased. A 3-pole stagger-tuned function, showing three tuned IF stages (A). The overall response of A is demonstrated by curve B. Center pole Q decreased (C), and the outer pole Q increased (D).

bandwidth, group delay, and response skirt selectivity. Fig. 7-6 illustrates a 3-pole function in its various stages, made up of three single-tuned circuits to produce the overall response illustrated in Fig. 7-6B. The curves at Fig. 7-6A represent three single-tuned circuits, such as a stagger-tuned IF system, with sound and video carriers at the top of each outside peak. The center pole Q is decreased at Fig. 7-6C and this causes the center dip, while the Q is increased among the outer poles at Fig. 7-6D, so the curve response expands at these points.

Siwko then proceeds to illustrate video IF changes in gain by varying the input impedance of a common-base transistor (Fig. 7-7). He tells graphically, how the transistor beta changes relatively little with variations of emitter current, and that the emitter-to-base resistance increases almost linearly with rises in emitter current. So, he deduces, in the reverse bias operating range of this common-base transistor, its gm varies just about linearly with the emitter current. In an actual circuit, with a beta spread exceeding 4:1, the gm stays within 20 percent. The equation for gm is given by the simplified equality:

$$gm = \frac{A(gain)}{Z_L(load)}$$

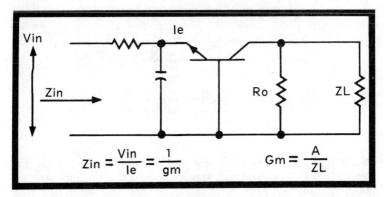

Fig. 7-7. Very simplified drawing of a common-base transistor IF stage. Gain is controlled by changing the input impedance.

He calculates the difference in useful gain between the common-base and common-emitter configurations to be 13 db.

In summary, Siwko remarks that the pole-shifting IF is relatively easy to design and suitable for mass production, with special importance attached to its use in integrated circuits. Obviously, such a stage or stages can easily be controlled by a well-designed AGC circuit, and probably will be, as developments continue in this very competitive consumer products race for bigger shares in the color television market. And automatic peak video gain under weak signal conditions, with the plus factor of noise reduction to boot, seems to make this type of IF-AGC approach quite attractive. This is another advanced circuit you will want to remember.

QUESTIONS

1. Vertical and horizontal sync pulses have but one purpose. What is it?

2. What was the original AGC detector and how is AGC operated today?

3. What are the usual plate characteristics of a sync separator?

4. Can the noise-gate adjust in Zenith's AGC-sync module affect AGC?

5. What's the usual noise pulse duration in any receiver?

6. Noise pulses last milliseconds or microseconds and usually affect how many scanning lines?

7. What is the significance of the IF curves in Fig. 7-5?

Chapter 8

Vertical Deflection System

In the past, vertical oscillator circuits have run the gamut from blocking oscillators, single-tube and common-cathode coupled multivibrators to circuits where the oscillator and output functions both were performed by a dual-purpose tube. Until the advent of solid state, an output transformer coupled the sweep signal to the yoke. Transistorized circuits use **capacitor coupling directly to the deflection yoke**. To top it off, there are now thick film vertical modules, and even an all-integrated circuit oscillator and output combination in a plastic package mounted on a heat sink.

The FCC vertical deflection specifications continue to remain the same, however. Each field must be scanned in 1/59.94 Hz, two fields to a frame, with 30 frames produced each second. The field scan time is 16.664 milliseconds, and the blanking interval, included in this count, is 1.4 milliseconds—the interval where broadcast sync and equalizing signals are inserted so they will not interfere with color and monochrome video.

Since vertical circuits operate quite close to house current frequency, they're not hard to build, but to design them with excellent linearity sans all sorts of control interaction is, to put it mildly, somewhat more than difficult. But we should see an end to such problems with thick film and even monolithic ICs where current generators are more commonplace. General Electric already has a monochrome receiver where separation has already been designed in at routine cost, according to engineer John Jordan.

VERTICAL OSCILLATOR-OUTPUT BLOCK DIAGRAM

As an example, let's consider a conventional (tube or transistor) vertical subsystem (Fig. 8-1) with the usual low-pass filter integrator network which blocks horizontal sync and shapes the vertical sync pulse. Vertical sync reaches the oscillator either in negative or positive polarity, depending on the circuit, to narrow or increase the conduction period of the

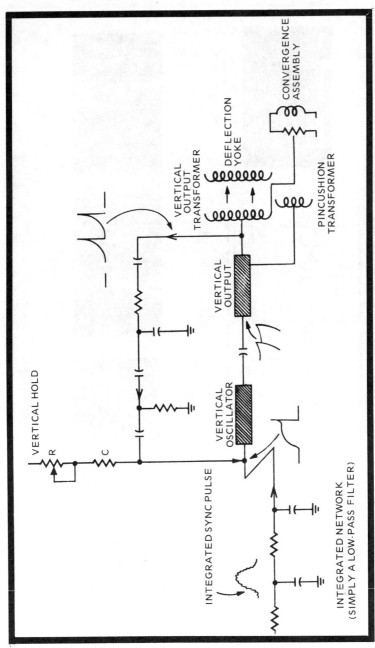

Fig. 8-1. Block diagram of a conventional vertical oscillator-output color subsystem.

197

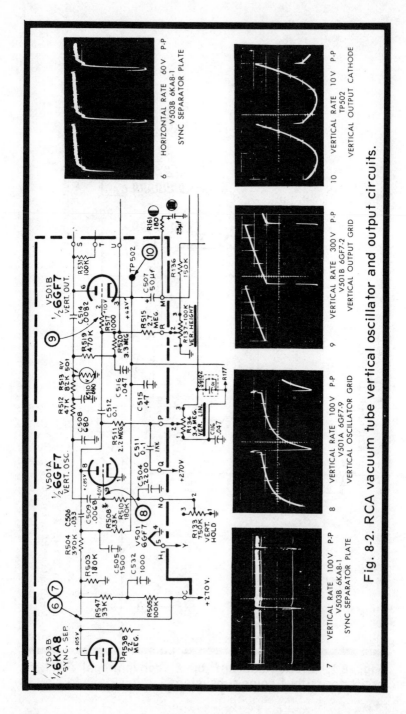

Fig. 8-2. RCA vacuum tube vertical oscillator and output circuits.

oscillator and, consequently, either speed it up or slow it down.

Tube-type oscillators deliver a trapezoidal waveform (a combination of pulse and sawtooth voltage) and transistor oscillators produce a simple sawtooth pulse. Usually, the output is cut off for the duration of the vertical blanking period denoted especially by the high overshoot portion of the vertical output waveform. From the output a portion of the cutoff voltage is fed back to the input of the vertical oscillator, developing an exponential type of voltage shown immediately after the integrated sync pulse, but not necessarily in proportion. When there is no incoming sync, R and C form a time constant that keeps the oscillator-output combination in flywheel operation so that a raster remains on the screen.

The vertical output supplies the vertical output transformer, and pincushion transformer.

Vacuum Tube Sweep System

Fig. 8-2 is typical of the vacuum tube circuits found in the remaining non-semiconductor receivers today. And even though the particular receiver we're discussing is a 1967 model, many other tube sets have duplicated much of the circuit because it's both simple and relatively reliable. Notice the 285-volt (and higher) plate voltages, the multitude of capacitors, and all the half-watt and larger resistors that must be added to develop the waveforms and time constants that make a vertical oscillator (V501A) and output (V501B) work together for a 1200 to 1500-volt output waveform. Compare this plethora of components and the operating potentials with those in the modular transistor and integrated circuits to follow!

In Fig. 8-2, sync from V503B (point 6) is applied to an integrating network (point 7) consisting of R504, C505 and C506 to the plate of the vertical oscillator and grid of the vertical output. Such a sync pulse could be strong enough to modify plate operation of the vertical oscillator, but especially the grid of the vertical output, since it is negative-going and can delay or speed up conduction of the output to time the receiver's oscillator. R508 is a voltage divider, while C504, C509, R510, and R133 form a filter and time constant network from the vertical output plate feedback which keeps the oscillator flywheeling at an approximate 59.95-Hz rate when there are no incoming sync pulses.

Vertical linearity control R134 varies the oscillator plate voltage which is drawn from the 1150 boosted-boost supply in

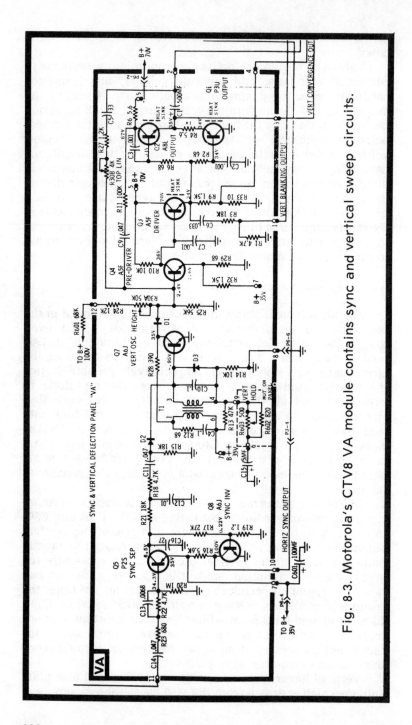

Fig. 8-3. Motorola's CTV8 VA module contains sync and vertical sweep circuits.

the high voltage. C511, R511, C512 and the voltage variation determine the linearity of the vertical trace during forward conduction. R520 and C516 are the main pulse-shaping components for the vertical output grid, to which is connected the DC height control from the 405-volt supply that determines the DC drive on the grid of the vertical output. C515 is simply a filter to pick up any transients from the suppressor grid of the horizontal output tube where the adjacent downward lead goes. Thereafter, 50-mfd filter C507 couples a 10-volt parabolic pulse from the vertical output cathode to the vertical section of the dynamic convergence assembly. The pulse is filtered by R161 and the 25-mfd capacitor.

The oscillator feedback network begins in the plate circuit of the vertical output. Feedback flows through C514 (DC blocking), a varistor (an AC sensitive resistor; when AC increases, resistance decreases), voltage divider R512-513, C508 and C510—integrating-type capacitors that shape the pulse so that it appears as shown in waveform 8. The waveform's sharp drop occurs when the tube drives suddenly into conduction, then the capacitors take over and permit an exponential curve until the oscillator cuts off, generating a small spike just prior to steep negative-going conduction. This tube does complete a full cycle swing since its DC plate is 285 volts and the peak-to-peak voltage shown at the grid of the output tube is 300v. The grid waveform, on the other hand, measures only 100 volts.

Here's an interesting point about that AC grid voltage: The exponential part of the capacitor discharge time goes up to DC (or zero) and only the spike goes positive enough to drive the oscillator tube into full conduction. The output stage drives the vertical output transformer, a pincushion correction transformer and vertical deflection yoke. In vacuum tube receivers, the inductive kick from the output transformer as the output stage turns off is several hundred extra volts—a good reason for RV501.

Motorola Modular Vertical Circuit

In Motorola's transistorized vertical circuit in the CTV8 series (Fig. 8-3), sync is coupled through C14 and the associated components to the base of the sync separator from the sync output stage in the collector of the second video amplifier. The emitter signal is swamped (filtered) by 100-mfd capacitor C601, but the collector signal is split two ways:

Voltage divider R17 and R19 provide bias for the Q8 sync inverter which reverses the phase of the horizontal sync pulse so that it will turn on the color gate pulse former on the adjacent panel (not shown) and also sync the horizontal oscillator. R15, R18, and R21, plus C11 and C12, form the capacitor charge-discharge integrating network for the vertical oscillator. While D2 is a guide and blocking diode that permits forward positive pulses but blocks negative ones. The blocking oscillator and its RC timing circuits are between the collector and base of vertical oscillator transistor Q7. As usual, the vertical hold is very much a bias circuit; C15 is part of the RC time constant with R603, and D3 is a limiting diode that will conduct and clamp when negative voltage swings produce sufficient negative bias. The height control is pretty much a voltage adjust for the collector of Q7 through guide diode D1 and, since it's purely resistive, the height should be quite linear. Top linearity control R30B is in the feedback loop between output terminal 2 and the base of pre-driver Q4 and is a series voltage adjust.

The primary and secondary of T1 are shunted by C4 and C10 so that the transformer is approximately resonant at 59.94 Hz. The frequency can be adjusted by varying the DC through the coil with the hold control and by the incoming frequency of the positive-going vertical sync pulse. Capacitors C4, C10 and C15 charge during the oscillator cutoff time through the B++ at terminal 7, and discharge rapidly through Q7 and perhaps R14. A blocking oscillator is used because of the better frequency stability over a wider temperature range; they're now small and less expensive, and greater linearity is possible with fewer external components.

When power is first applied, C4, C10 and C15, charge rapidly through the applied DC voltage until they reach a value that is positive enough for the oscillator to conduct. Since Q7 is simply a transistor switch, its cycle shorts to ground, discharging the capacitors and, in so doing, drops the forward base voltage enough to cut itself off. The capacitors then charge all over again, bring the base to the conduction level, once more biasing the vertical oscillator into conduction. The DC divider action of R603, along with incremental discharges of C15, set the timing for the oscillator.

As Q7 becomes a 16.664-millisecond switch, it causes conduction in the emitter forward biased pre-driver (an inverter) that turns on the driver. The driver, via an emitter connection, supplies vertical blanking, positive vertical pulses for the convergence panel (not shown), and switching for the complementary vertical output. Since Q2 is an NPN type and Q1 a PNP type, the negative and positive gating from the driver turns on first one and then the other for a low-impedance push-pull output that is used for both vertical convergence and to drive the vertical deflection coils with a linear sawtooth deflection **current**. Capacitors C2 and C3, shunting each transistor base-to-collector, are low-pass filters called wave-shapers and transient suppressors. The output reactance of C1 (500 mfd), by the way, is 5.33 ohms at 59.94 Hz, a rather low impedance.

Troubleshooting some of these DC-coupled stages can be quite another matter, as an illustrative analysis will indicate. Suppose, for instance, the vertical oscillator quit for some reason and the height control was turned to a midpoint value of 25K. The voltage at the base of pre-driver Q4 would become 39 volts, driving this transistor into saturation; its collector voltage would drop well below the 34 volts at the emitter of Q3, cutting Q3 off. As a result, there would be no output. If Q4 shorted, the collector voltage would fall between R10 and R29, Q3 would be biased hard off, and the absence of emitter voltage would turn off Q1 and Q2. Again, there would still be no output. An open Q3 would produce the reverse conditions since R10 would allow something less than a full 70 volts from B + across it to the base of Q3, driving Q3 hard on, and turning Q2 on also. If the driving voltage became sufficient, it might even burn up Q2, depending on the emitter circuit impedances. A shorted Q3 would do the same overdrive thing, while an open Q3 would remove bias and cut both Q1 and Q2 completely off— all very good reasons for looking at AC and DC voltages together with your oscilloscope when repairing or analyzing either video or sweep circuits anywhere in any television receiver.

MOTOROLA IC VERTICAL DEFLECTION SYSTEM

The designer of this all-integrated circuit deflection system (Fig. 8-4) was Milton Wilcox of Motorola's Phoenix semiconductor facility, recipient of the Broadcast & Receiver IEEE Group's best circuit award in 1971. In it he does a number of unusual things:

1. Eliminates the feedback loop from the output to the oscillator, which inevitably results in control interaction and affects interlace.

2. Eliminates transformer output coupling, thereby having the drive circuits deliver a restricted voltage across the scan (yoke) coils for a linear sawtooth current, with the current dependent on the coils' impedance.

3. Gives the output circuit an open-loop output impedance to avoid including the usual thermistor and linearity control.

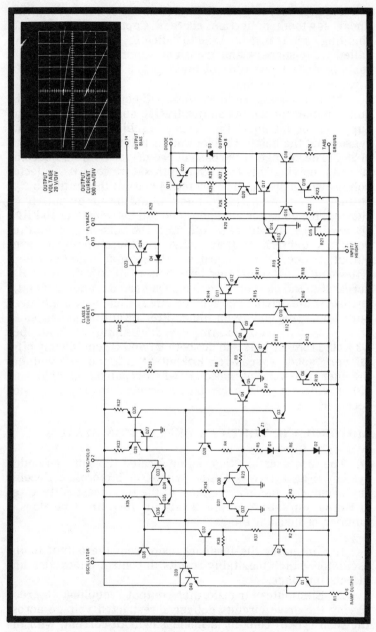

Fig. 8-4. Schematic of the Motorola Phoenix vertical output subsystem on a single IC chip. Waveforms show voltage and current outputs. (Courtesy Ben Scott)

4. Includes a flyback generator to supply a fixed voltage pulse that also can produce vertical blanking.

5. Puts the entire assembly on a monolithic chip in an in-line plastic 675 package, 14 pins (4 used by the heatsink), 0.725 inch long by 0.255 inch wide.

The schematic, with output voltage and current waveforms, is shown in Fig. 8-4. The sawtooth oscillator is biased by Z1, Q28, D1, and D2, and has a single external capacitor (not shown) used for timing connected between pin three and ground. During scan time, this capacitor is charged by current source Q25 to the point that differential amplifier Q33 through Q36 is switched by an incoming negative sync pulse at the sync-hold terminal, pin 2. This latches Q37 and Q38, turning on Q6, while turning off Q5 and Q7. NPN current source Q3 discharges the external capacitor, forming a sawtooth voltage. This is the flyback part of the IC operation. Vertical scan commences again as differential amplifier Q29 through Q32 switches, turning on Q2 so that it, in turn, shuts off Q37 and Q38. The output is delivered through Darlington stages Q39 and Q40 at the collector of Q1 as a ramp (or sawtooth). The positive ramp from the external capacitor is used during forward trace, while a pulse generated during retrace controls the flyback.

During flyback, a capacitor connected between pins 9 and 12 is charged by the Q8-Q9 current source through diode D4, an external diode from pin 9 to the power supply. When the cycle begins, Q8 and Q9 are off, and Q15 plus Q19 clamp the bases of Q16 through Q18 to ground. Peak current now flows out of the yoke to the supply through D3, D4 and Q23. As the yoke current reverses, Q23 and Q24 clamp the negative side of another external capacitor (not shown) to the supply voltage, while Q21 and Q22 saturate with the coil current increase. During flyback, D4 (external) is reverse biased and so a flyback voltage higher than the supply voltage develops at the scan coils. This is the boosted voltage for the output.

The output circuitry consists of driver Q13, Q14, Q16, and Darlington outputs Q21, Q22, Q17 and Q18. An oscillator sawtooth is applied at the input-height connection pin 7 (Fig. 8-4, a 90-degree color application), sending Q13, Q14, Q17 and the associated transistors above into conduction. At the beginning of scan, Q16 through Q18 are turned off, and Q21 and Q22 are fully on, delivering peak current across R28 because of the bias delivered by R26, R27 and R29 from the output bias at pin 14. Meanwhile, the input quiescent voltage is set by resistors R14 through R16, at the base of Q11, Q12, then to Q13 and Q14 by way of a second divider, R17 and R18. A voltage is then

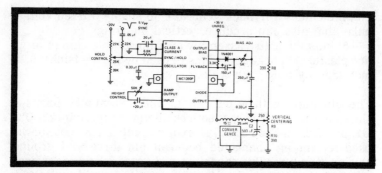

Fig. 8-5. Application drawing of Motorola's MC1390P 90-degree vertical deflection module.

developed across R24 in the Q18 output that sets the Class A quiescent current shown at pin 1 with a 6.8K resistor to ground and a 20-mfd filter. If there are unwanted variations, Q10 conducts and regulates any excess current, while an external resistor between diode pin 9 and output bias pin 14 determines the Class A current in Q21 and Q22. Q20 is a short circuit protection transistor across the top half of the output.

As scan continues, Q17 and Q18 begin to conduct, lessening the output current. Q16 also becomes active, shunting current from the upper bias network, diminishing its flow through the bias resistors, and reducing IR drops. Therefore, Q21 and Q22 produce less output. Current now is increased through Q17 and Q18 until, at the center of forward scan, currents equal the Class A current and no output current flows.

With further scan, Q21 and Q22 are gradually brought toward cutoff, while Q17 and Q18 conduct harder until, at the end of the scan, these two transistors are actually removing Ipeak due to voltage developed across R24. However, Q21 still follows the lower group of output transistors, and so continues to offer a high output impedance. Ipeak must flow through the upper stages at the beginning of the scan and through the lower stages at the end of the scan. For higher deflection currents (110 degree tubes) and low-impedance scan coils, two XC1390 ICs may be placed in parallel to deliver scan currents up to 2 amps p-p.

ZENITH'S DISCRETE & THICK FILM VERTICAL DEFLECTION

The original Zenith vertical deflection subsystem circuit for the solid-state 40BC50 is illustrated in Fig. 8-6. Sync inverter Q705 inverts the negative-going sync signal so that it turns on NPN gate Q701 during the sync pulse interval, which

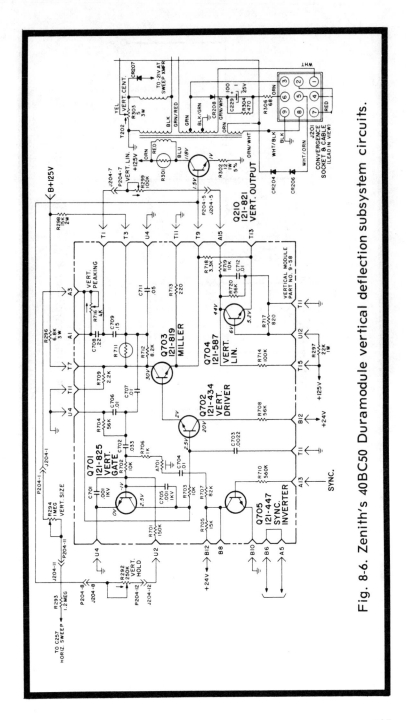

Fig. 8-6. Zenith's 40BC50 Duramodule vertical deflection subsystem circuits.

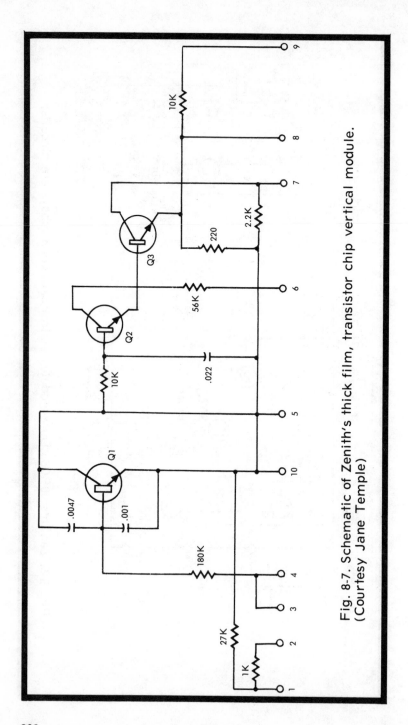

Fig. 8-7. Schematic of Zenith's thick film, transistor chip vertical module. (Courtesy Jane Temple)

lasts for the 1.4-millisecond vertical blanking interval. Collector potential for Q701 and the base drive for Q702 are supplied through vertical size potentiometer R294, with vertical peaking connected to this same line and also to capacitor C708 in the collector of Q703, the Miller amplifier. And connected between C708 and C709 is one end of the vertical linearity control; the other end goes to the collector of vertical linearity Q704 to shape the output during scan time for the best rising forward trace.

The vertical size potentiometer's source voltage comes from a rectified pulse from the flyback transformer, so that any change in the pulse amplitude will affect the vertical size and cause it to track with the horizontal. Q702 also receives drive from the sawtooth developed by the Miller amplifier through C708 and the vertical peaking control. Q702 is DC emitter-coupled in phase to the base of the Miller amplifier. Output of the Miller stage is emitter-coupled to the base of the vertical output. Another output from collector feeds the base of Q701 through C707, C702 and R702. Also, the Q703 output takes its Miller route through C708 and C709 to the base of the vertical driver. The thermistor takes care of temperature changes by proportionally increasing or decreasing in resistance.

As the receiver is turned on, C702, C706 and C707 are charged through several paths from B+, while Miller capacitors C708 and C709 wait for charges that take place only during retrace. This is because the Miller oscillator is not conducting at turn on and the collector voltage goes to almost full DC—the point where C709 and C708 take on full charges. With Q701 in saturation, the charge on C702 is dissipated, and Q701 comes out of conduction. Capacitors C708 and C709 then take over and begin to discharge during forward trace time, turning on Miller amplifier Q703, which further discharges C708 and C709. The positive sawtooth present in the emitter of Q703 is applied through the vertical hold to the base of Q701, and the vertical hold and C702 form a time constant that both restarts the entire process and also keeps the oscillator operating when there is no incoming sync pulse. Vertical peaking, naturally, affects the discharge waveform (undoubtedly, its symmetry), while the vertical linearity transistor is something of an active circuit that could change with variations in DC, preserving sweep symmetry by a second method.

There is another version of the same circuit that includes most of what is shown in Fig. 8-6, except sync inverter Q705 and vertical linearity transistor Q704. In Fig. 8-7, the linearity control is connected directly to the emitter of Miller Q3 through a 10K resistor. There are a couple of other minor changes, but the big point of this illustration is that the revised circuit is also the new thick film transistor-chip module. The final number in the 700 series now corresponds to "our" numbers on the thick film diagram. The 9-751 and the new thick film circuit should be direct replacements. The difference, of course, is that the heart of the new circuit is in a sealed ceramic package and must be replaced in its entirety should there be a fault.

QUESTIONS

1. Name the field scanning rate of the vertical oscillator. How many fields to a frame? What is the field scan and retrace time?

2. What's the biggest problem in the vertical circuits we're discussing?

3. A tube-type vertical oscillator delivers what kind of a waveform to the vertical output? The transistor oscillator waveform?

4. Pulse-shaping networks in vertical oscillator and output circuits are always what kind of components?

5. Why does the vertical oscillator stay in virtual sync and continue to drive the vertical output when there is no incoming sync signal?

6. What are the advantages of blocking oscillators?

7. What would you expect the X_c reaction of a vertical output capacitor to be?

8. How can the use of a single IC help you troubleshoot an entire vertical subsystem?

9. The two most active capacitors in Zenith's Miller amplifier are?

Chapter 9

Horizontal Deflection System

AFC diodes and the associated circuits have been tough cookies in one way or another for a long time, and designers approach new applications with more than a little caution. For instance, in working with AFC diodes, you must be aware of the circuit's hold and pull-in ranges, residual phase differences, noise-bandwidth, pull-in time, and damping factor, plus other lesser known considerations. One of these is the interaction between AFC and AGC systems, especially when the AGC time constant is reduced to allow faster IF bias response. This decreases the horizontal AFC's pull-in range, increases the video IF response and detector output, and so increases ripple voltage output. Sync tracks the ripple only where it increases sync amplitude and, since it will not track in the opposite direction, the sync train to the AFC diodes is incomplete, and this decreases the loop gain of the automatic frequency control system, retarding its efficiency and causing possible horizontal sync-hold problems.

The ripple voltage mentioned is produced during the short-period coincidence of the broadcast sync signal and flyback feedback to the AFC diodes. At that point, the tuner and IF stage gain is reduced, which lowers the video input to the AGC system. As the cycle repeats, a ripple voltage develops that is the frequency difference between the incoming 15.734-kHz horizontal sync pulse and the AGC-RF-IF loop relation.

There are other problems in horizontal circuits such as the design of transistorized horizontal output amplifiers, the switching time, secondary breakdown, power handling, and thermal (temperature) stability. Saturation losses in such transistor amplifiers occur during forward scan periods, and will increase rapidly with higher temperatures. But the largest loss is at fall or turn off time when high voltages and currents coincide, causing the highest collector dissipation. Obviously, heat sinking to take care of this dissipation is more important, and the transistor(s) must have good contact (often silicon grease) to maintain adequate thermal

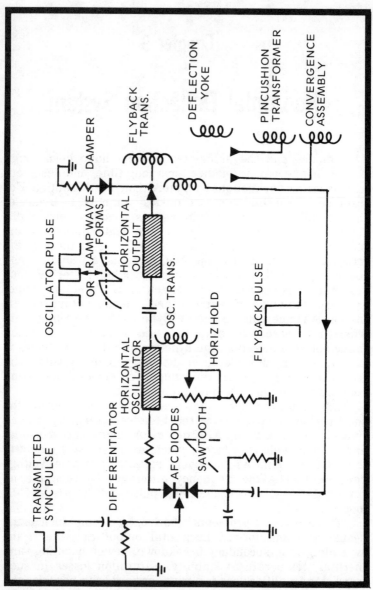

Fig. 9-1. Block diagram of a typical horizontal AFC, oscillator, output and feedback line-scanning subsystem. The forward operating time is 52.4 microseconds, retrace 11.1 microseconds—a total of 63.5 microseconds. Dampers are either single diodes in transistor-tube circuits, or don't exist in dual SCR systems.

dissipation. Precisely the same considerations apply to silicon controlled rectifiers, such as those used in pairs by RCA for forward trace and retrace drive times. And, we might note, if semiconductor horizontal output stages are driven too fast by a runaway oscillator, they will heat somewhat proportionally to the cyclic rate, quickly turn red, and silently fade away, leaving no high voltage and an interesting repair problem for the service shop.

HORIZONTAL BLOCK DIAGRAM

The horizontal block diagram in Fig. 9-1 is somewhat reminiscent of the vertical circuits, except that feedback and sync comparison circuits are more elaborate, and the entire system operates at about 263 times the vertical rate 59.994 Hz x 263 equals 15,734 Hz, or thereabouts at slide rule accuracies.

Differentiated broadcast sync and flyback sawtooth feedback pulses meet at the AFC diodes, which by conducting relatively positively or negatively on resolved phase—and-or frequency differences, deliver a plus or minus DC correction voltage to the horizontal oscillator to keep it on frequency. Some oscillators have blocking transformers, others are RC coupled, and still a third group features sine-wave output coils that are tunable for the correct sine-wave symmetry. Many horizontal hold circuits are standard variable resistances, except Zenith, for instance, where the hold control is the tunable oscillator coil.

In semiconductor circuits, the oscillator drive and output waveforms are pulses, while in vacuum tube receivers the drive voltage for the output is a ramp-type exponential waveform that turns on the output tube at about ¾ths of the waveform peak. The output conducts at a horizontal rate, firing the flyback to drive the deflection yoke as well as producing additional voltages for the pincushion transformer and convergence assembly. The damper, if there is one, conducts after the horizontal blanking interval. The collapsing magnetic field during retrace snaps the horizontal sweep back to the left side of the tube and then helps move the trace on its way toward center. Forward scan, of course, propels the trace from the center to the right edge of the CRT, completing the forward sweep.

The damper has a basic function other than helping to move the trace, however, and this is doing what its name implies. The inductance of the horizontal output transformer and the deflection yoke become a resonant circuit with the internal distributed capacitance. When forward trace has

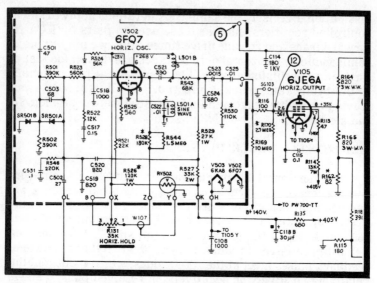

Fig. 9-2. Schematic of an RCA vacuum tube horizontal oscillator-output circuit used in 1966-1967.

ceased, an oscillation begins because of shock from magnetic reversal of the system and the generated high voltage. **Were this oscillation permitted to continue for any period, there would be a raster foldover on the left side of the screen.** To keep this from happening, a tube or semiconductor diode is inserted in the circuit to collapse the magnetic field rapidly and, at the same time, heavily load the flyback and entire output circuit, effectively damping any visible oscillation. Small residual oscillation has sometimes been made use of to tune the horizontal output transformer through a coil to either the third or fifth harmonic for maximum efficiency. Most yoke-flyback systems, however, including the newest ones, remain untuned, since distributed capacitance and internal inductance usually peak these circuits at about the third harmonic without the necessity of an external means of controlling the internal resonance.

VACUUM TUBE HORIZONTAL SWEEP CIRCUIT

At first, the 6FQ7 in Fig. 9-2 looks very much like the original pulse-width modulation system RCA used when grandaddy double-triode 6SN7s performed the control and oscillator function. The appearance is somewhat the same, but the action is different. The two opposing diodes receive both sync and flyback pulses for phase and frequency comparison, and a DC-AC correction voltage is transferred through R523 to the grid of the control tube. This generates plate current through pin 1 that is passed to the center of the horizontal hold control. Out of hold control terminal 3, current can flow back to the cathode and-or result in a cathode-plate current mix, depending on incoming AC or the setting of the horizontal hold control. This control opens at one end when it closes at the other, depending on the feed to the grid of the actual oscillator portion of the tube.

Interestingly enough, there is a 1.2-volt sawtooth waveform at the grid of V502A that is inverted and amplified, since the cathode is degeneratively biased and the plate receives its potential through a varistor (voltage-sensitive resistor) from the +405-volt supply. Composite pulses from the plate and cathode combine in the horizontal hold resistance and are attenuated more or less by R131. Depending on the AC amplitude and DC levels, the pulses either speed up or slow down the oscillator to keep it in step with the transmitted sync signal.

Sine-wave oscillations are generated in the tank circuit composed of C522 and L501A, with oscillator feedback through L501B. The V502B plate voltage comes from the 405-volt source, while C524 and C523 are AC output shaping and voltage-dividing capacitors along with R530. Output tube grid leak bias is developed by C525 and 10-meg R169. At pin 2 of the oscillator, R522 and C517 combine as a filter, called an anti-hunt network, that keeps the oscillator from "hunting" (slight changes in frequency) rapidly and overcorrecting when the receiver is initially turned on. From C525, an exponential sawtooth is applied to the grid of the horizontal output through oscillation suppressor R116. Spark gap 103 provides extraordinary high pulse protection.

During grid leak, when the control grid of V105 is negative because of C525's discharge, the horizontal output tube is cut off and the horizontal output transformer reverses its magnetic field, producing a large pulse voltage that is processed for high voltage. As the exponential 200-volt grid drive waveform reaches about ¾ths of its peak, the horizontal output grid is positive enough so that it turns on V105, a power amplifier, which drives the flyback transformer and associated high-voltage circuits, including the sweep. The cathode of the output tube is directly grounded, the screen is at a positive 145-volt potential, while the suppressor is at a low DC potential to keep secondary electrons from bouncing back from plate to screen.

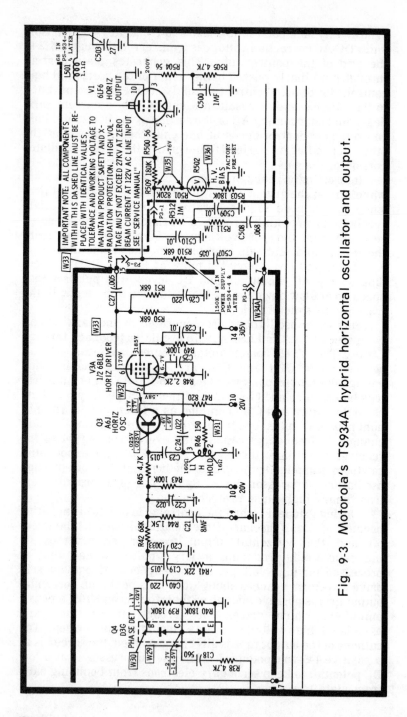

Fig. 9-3. Motorola's TS934A hybrid horizontal oscillator and output.

Later horizontal tube circuits have an output tube regulating device that is basically a pulse (and-or DC) feed from the flyback transformer through a capacitor and varistor that biases the output tube grid to keep it from going heavily positive and burning up the plate circuit should there be too heavy a high-voltage demand.

In troubleshooting this circuit, look at the oscillator frequency first. If it is working, see if its drive rate is close to 15.7 kHz; your scope can easily tell you. If the oscillator is operating on frequency, go to the rest of the circuit and look for problems there. Unfortunately, many potentials in this circuit after the 268-volt plate of the oscillator will permit the use of either AC or DC oscilloscope vertical amplifiers, but not both. So you will have to switch them manually and alternate between the two to establish AC amplitudes and waveshapes, then investigate DC potentials.

HYBRID HORIZONTAL SWEEP CIRCUITS

Hybrid horizontal sweep circuits are much like the block system shown in Fig. 9-1, and considerably simpler than the usual all-vacuum tube circuits, since AFC diodes easily supply effective correction to transistor single-stage oscillators—especially with vacuum tube drivers to supply the large drives needed for the output. Why is this method used instead of all solid-state oscillators and outputs? Simply because the manufacturing process is cheaper. Does a tube horizontal output give better results? It will if the transistor output circuit doesn't have some sort of eventual regulation. The transistor isn't husky (with a good fudge factor), and there must be some means of suppressing transients. On the other hand, semiconductors are always going to last longer than tubes in most applications; therefore, they are undoubtedly more desirable.

You might notice in Fig. 9-3 the protection system using high-voltage bias (varistor R502) and the voltage divider in the grid circuit of the 6LF6 horizontal output. This is the HV pulse-type output protection and limiting circuit we've been talking about. Regulating pulses arrive between R501 and R502 from the primary of the flyback and voltage-dependent resistor R502 increases and decreases in resistance, depending on the minimum and maximum amplitude of the flyback pulses. The R502 resistance change regulates the grid conduction of the horizontal output, so excessive current is not drawn because of beam current demand. The height of the

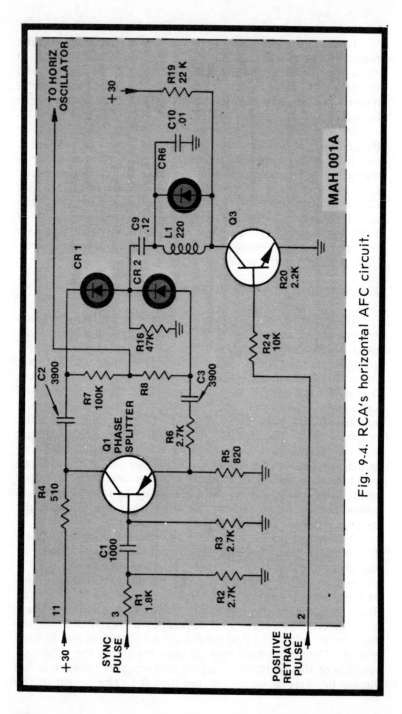

Fig. 9-4. RCA's horizontal AFC circuit.

pulse is, of course, proportional to the amount of high voltage. R503, the HV bias adjust, is set at the factory and normally does not require changing, but should be checked after oscillator, HV, or horizontal output repairs or replacement. In some other versions of this circuit, a pair of diodes is added to supply a negative cutoff potential to the output grid, insuring the tube's safety whenever the oscillator fails.

RCA Silicon Controlled Rectifier Horizontal Drive System

Although this system is several years old, it will be an RCA mainstay for a considerable period in the future. Really, all RCA 1972 solid-state chassis have this basic system, plus the CTC40, 44, and 47 receivers that preceded the CTC49, and the series of CTC54 and 46 chassis. It's a good system, a reliable one, and fundamentally simple, once you understand its operation; but understand it you must, and this sometimes takes a bit of doing. The following description is based on studies produced by Carl Moeller and Ed Milbourn, RCA senior Field Technical Services Administrators.

The horizontal AFC and oscillator circuits are contained in Module MAH. The AFC circuit is shown in Fig. 9-4. Differentiation of the combined vertical and horizontal sync pulses by C1 and R3 eliminates the former; the horizontal sync pulses are split in phase and supplied to the two inputs to the phase detector, C2 and C3. Positive retrace pulses from the flyback transformer are shaped in the collector circuit of Q3 before being applied to the common point of CR1 and CR2.

Assuming the free-running frequency of the horizontal oscillator is exactly equal to sync-pulse frequency and correctly phased, the output of the AFC circuit is zero. A tendency of the horizontal oscillator to lag the sync pulses produces a positive AFC output and, of course, if the oscillator output pulse tends to precede the sync pulse excessively, a negative AFC voltage is generated.

The horizontal blocking oscillator itself is diagrammed in Fig. 9-5. Fig. 9-9 is a complete schematic of the horizontal output system, excluding the high-voltage quadrupler.

There are three resonant circuits to be considered in the operation of the yoke circuit. These are:

1. The trace resonant circuit consisting primarily of C408 and C409 in parallel. T404, the yoke, L404, and either CR101 or SCR101, whichever is conducting. The yoke inductance is by far the greatest inductance, comprising roughly 90 percent of the total. The resonant frequency of this circuit is **roughly** 10

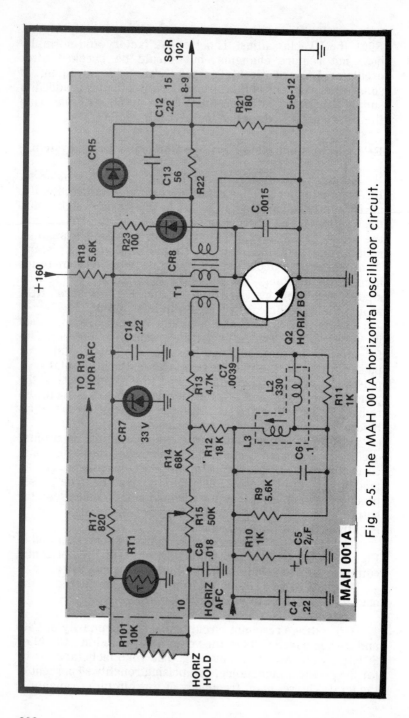

Fig. 9-5. The MAH 001A horizontal oscillator circuit.

kHz; therefore, the period of one-half cycle is approximately equal to the duration of one scanning interval. Refer to Fig. 9-6.

    2. The retrace resonant circuit made up of C408 and C409 in parallel, T404, the yoke, L404, C413 and C414 in parallel, L104, and either CR102 or SCR102, whichever is conducting. The resonant frequency of this circuit is such that its half-cycle period is equal to retrace time. Refer to Fig. 9-7

    3. The power input resonant circuit consisting of R424 and T402 in parallel with T403, L104, and two parallel paths to ground. One of these is C412 and R407 paralleled by L406; the other is C413 and C414 in series with either SCR101 or CR101, whichever is conducting. The resonant frequency of this circuit allows the voltage at the junction of T403 and L104 to rise to its positive peak and begin to decay during each scanning interval. The precise frequency is varied by the

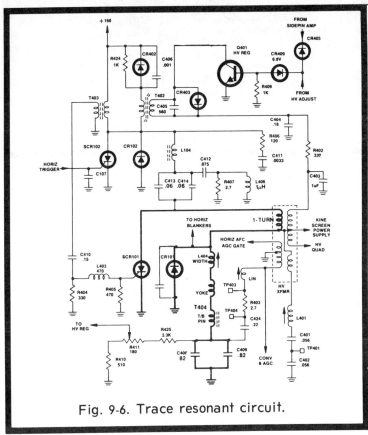

Fig. 9-6. Trace resonant circuit.

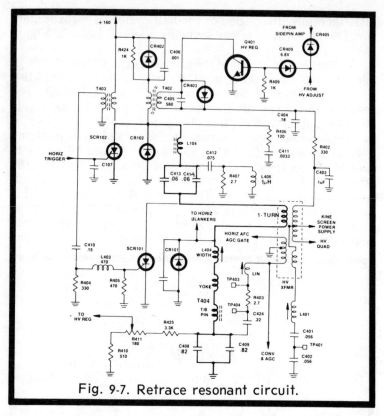

Fig. 9-7. Retrace resonant circuit.

voltage regulator, to be explained. Current flowing through T403 generates a positive gate voltage for SCR101 during most of the scanning interval. Refer to Fig. 9-8.

Considering a point in time just after retrace is completed, time 0 in Fig. 9-10, a large current is flowing in the trace circuit and conduction is through CR101. This current decays to zero at midscan and reverses, flowing through SCR101 and rising from zero to maximum at the right end of scan. Near the end of the scanning interval, gate voltage is removed from SCR101, but it continues conducting.

Throughout the scan interval, the lower ends of C413 and C414 effectively are grounded. Electron current flows from ground through C413, C414, C412 and the rest of the power input circuit to B+. L104 is small compared to T403 and T402, so the voltages at its opposite ends are about the same. The voltage at the anode of SCR102 and CR102 reaches maximum slightly before the end of trace and begins to decay, since the power

222

input circuit is resonant. Regulation of input power is accomplished by changing the resonant frequency; this is the function of the saturable reactor, T402.

When SCR102 is triggered, the retrace resonant circuit is completed and the energy stored in C412, C413, and C414 causes electron flow to continue downward through the yoke, rapidly cutting off SCR101.

Resonance causes the yoke current to decay to zero and reverse, cutting off SCR102 and turning on CR102. At this instant, scan is at the center of retrace. Current increases until it reaches a maximum, driving the scan to the left side of the screen, but it cannot reverse again because gate voltage no longer is present on SCR102. The large current flowing in the yoke at the end of retrace takes the only path available, through CR101. This, of course, was the condition assumed at the beginning of this summary.

The high-voltage transformer is connected parallel to the yoke circuit. During scan time either SCR101 or CR101 is

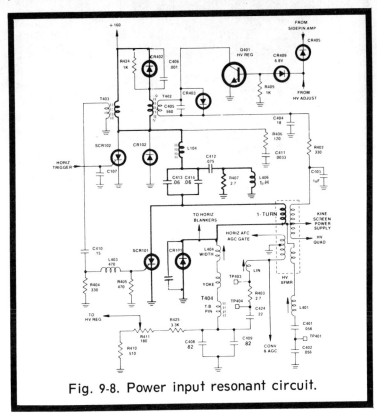

Fig. 9-8. Power input resonant circuit.

223

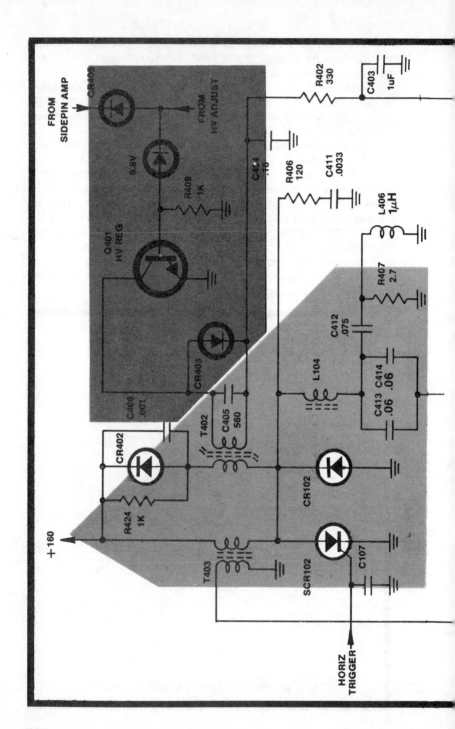

224

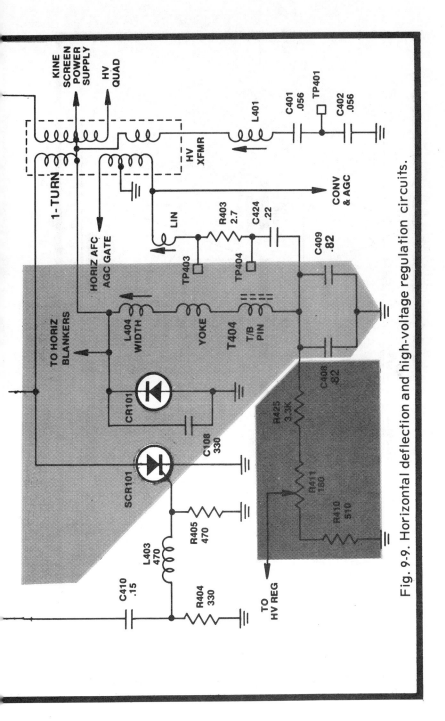

Fig. 9-9. Horizontal deflection and high-voltage regulation circuits.

225

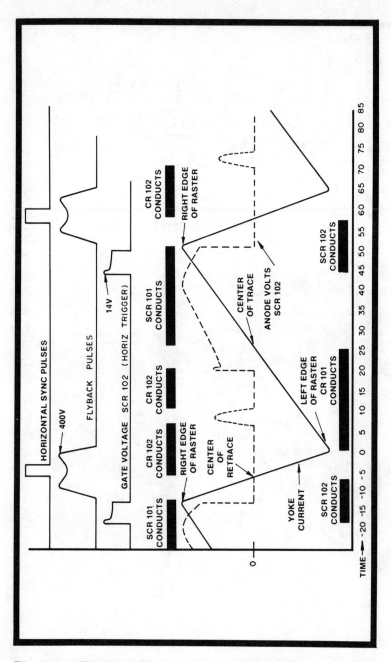

Fig. 9-10. Timing diagram of the RCA SCR horizontal deflection system.

conducting, thus shorting the primary so that very little energy can enter. However, during retrace the flyback voltage pulse drives the transformer primary, and the transformer converts this voltage to a higher voltage, which is used to drive the quadrupler and CRT screen power supply. Energy transfer is optimized by tuning the transformer to the third harmonic of the retrace pulse. This is the function of L401.

The Basic SCR System

A silicon controlled rectifier (SCR) is—as its name implies—a controlled rectifier fabricated from the semiconductor material, silicon. The SCR action is much like that of the thyraton tube. It is basically nonconductive until turned on by some control signal applied to a control electrode. When turned on, the device acts like a normal rectifier, and is capable of conducting high currents and exhibiting very low forward voltage drop between its anode and cathode. The SCR can be turned off by reversing the anode-cathode voltage or by reducing the current through it to the point where the SCR exhibits a high resistance.

The SCR has three elements—an anode, a cathode and the control electrode called a gate (Fig. 9-11). When a sufficient amount of current is supplied to the gate, and the anode-cathode is forward biased, the SCR will switch on. When the anode-cathode voltage and-or current are reversed, the SCR will switch off. The amount of forward anode-cathode voltage which will result in SCR conduction depends on the gate current (Fig. 9-12). After the SCR has been switched-on, the

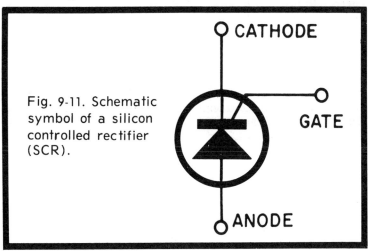

Fig. 9-11. Schematic symbol of a silicon controlled rectifier (SCR).

CATHODE

GATE

ANODE

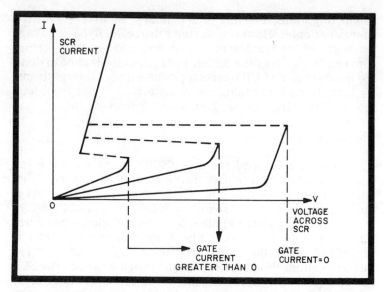

Fig. 9-12. SCR conduction and gate current relationship.

current through it is independent of gate current. The fact that the SCR exhibits the characteristics of a switch is used to aid in the upcoming simplified description of horizontal deflection circuit operation.

The basic objective of any horizontal deflection circuit for commercial television receivers using electromagnetic deflection is the same—to cause an approximately linear current to flow in the yoke windings in such a manner as to deflect the picture tube beam from one side of the tube screen to the other (trace current). This current must be in synchronism with the video information provided in the received television signal transmission. Additionally, the yoke current must include a component which causes the picture tube beam to return to its starting position; this is, of course, the retrace or flyback current component. The basic simplified horizontal deflection waveshape and its relationship to beam scan is illustrated in Fig. 9-13.

The CTC40 chassis generates the desired horizontal deflection current with circuitry using silicon controlled rectifier devices and associated circuit elements. The essential components in the RCA CTC40 horizontal output circuit are as shown in Fig. 9-14. Essentially, diode D1 and controlled rectifier SCR1 provide switching action which controls current in the horizontal yoke windings, L_y, during the picture tube beam trace interval. Diode D2 and controlled rectifier SCR2 control yoke current during the retrace interval. Components L_r, C_r, C_h, and C_y provide the necessary energy storage and timing duties. Inductance Lg_1 supplies a charge path for C_r and C_h from B+, thereby providing a means to "recharge" the system from the instrument power supply. Inductance Lg_2 provides a gating current for rectifier SCR1 (Lg_1 and Lg_2 comprise transformer T102). Capacitor C_h optimizes the retrace time by virtue of its resonant action with L_r.

In order to facilitate the understanding of the operation of the horizontal output circuit, liberty has been taken to greatly simplify the circuit element illustrations in the following discussion. In Fig. 9-15 (and the illustrations following), trace elements SCR1 and D1 together comprise an SPST switch

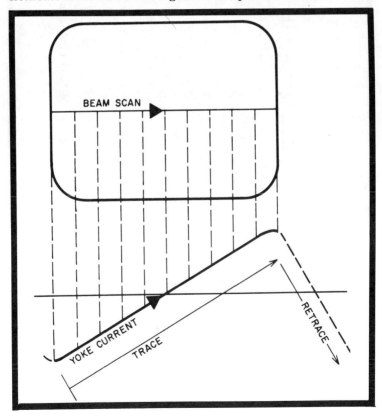

Fig. 9-13. Basic relationship between beam scan and yoke current.

229

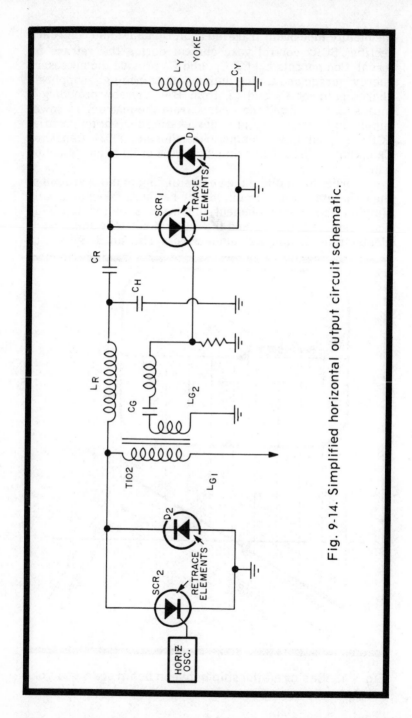

Fig. 9-14. Simplified horizontal output circuit schematic.

labeled S1. Retrace elements SCR2 and D2 comprise another SPST switch labeled S2. The relationship of the yoke current waveshape and the beam trace-retrace also is illustrated in Fig. 9-15.

Trace time: Referring to Fig. 9-16, during the first half of trace time (T_0 to T_2), switch S1 is closed, allowing a field previously produced about the yoke inductance, L_y, to collapse, resulting in a current which charges capacitor Cy. This yoke current causes the picture tube beam to deflect to approximately the middle of the screen. During the second half of the trace time interval T_2 to T_5, the current in the yoke circuit reverses because the capacitor C_y now discharges back into the yoke inductance, Ly (Fig. 9-17). This current causes the picture tube beam to complete its trace.

Retrace initiation: However, at time T_3 (Fig. 9-18) a pulse from the horizontal oscillator causes switch S_2 to close, releasing a previously stored charge on C_r. The resulting current flows in the circuit as illustrated in Fig. 9-18. Because of the natural resonance of L_r and C_r, this resonant current

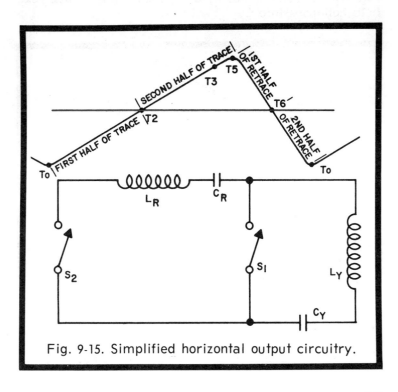

Fig. 9-15. Simplified horizontal output circuitry.

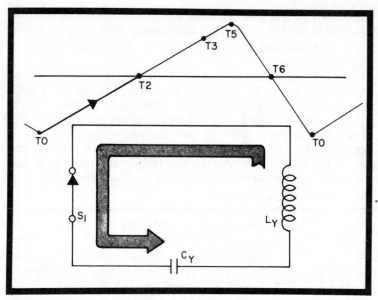

Fig. 9-16. Circuit current T0 to T2; the first half trace yoke field collapses into Cy.

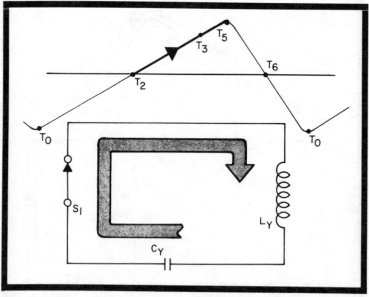

Fig. 9-17. Circuit current T2 to T5; during the second half of the trace, Cy discharges into the yoke windings.

becomes equal to the yoke current at T_5. At this time S_1 opens, S_2 remains closed, and retrace starts.

Retrace: The simplified retrace circuit with S_2 closed and S_1 open is shown in Fig. 9-19. Basically, L_r, C_r, C_y, and L_y are connected in series. The natural resonant frequency of this circuit is much greater than that of the yoke circuit, L_y and C_y. As a result, the current change through the yoke windings L_y during retrace time is much faster than that during trace time. This, of course, causes the picture tube beam to retrace (flyback) very rapidly.

During the first half of the retrace time (T_5 to T_6, Fig. 9-19), the retrace yoke current flows in the direction as shown at time T_6; however, the retrace circuit current reverses because of resonant circuit action (Fig. 9-20), and the last half of the retrace action occurs (T_6 to T_0).

At time T_0, switch S_1 closes and shortly thereafter S_2 opens. The field about the yoke inductance starts to collapse, and the resulting current again starts the trace interval. The trace-retrace cycle now has been completed. Fig. 9-21 illustrates the trace and retrace waveforms.

ZENITH'S CERAMIC THICK FILM MODULE

Zenith is not only moving rapidly forward with monolithic ICs, but also with thick film complex circuits—sometimes

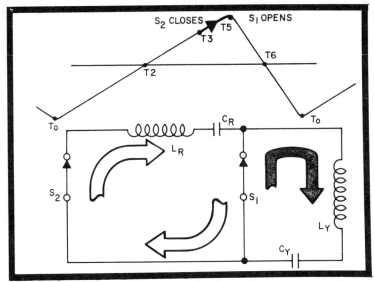

Fig. 9-18. Circuit current (T2 to T5) during retrace initiation.

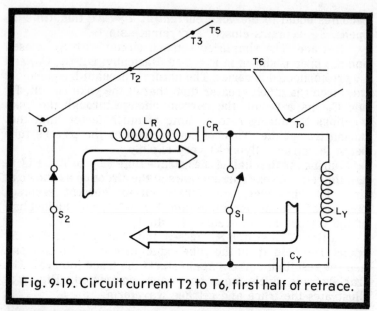

Fig. 9-19. Circuit current T2 to T6, first half of retrace.

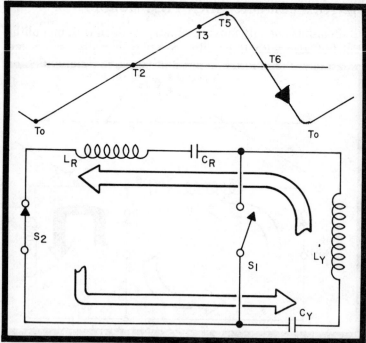

Fig. 9-20. Circuit current T6 to T0, second half of retrace.

intermixed, as shown in Fig. 9-22. Here, we have a thick film resistor matrix module for the chroma subcarrier, a thick film vertical module, and the horizontal AFC and oscillator module. What's happening in these circuits is that Zenith is more or less substituting transistor chips and thick film modules for the individual transistors and discrete passive components that formerly constituted either part of or the entire module. Fig. 9-23 is a schematic of the original horizontal Duramodule used in the 40BC50, Zenith's first all-solid state receiver, while Fig. 9-24 is the schematic of the actual thick film circuit that supplants it. Don't be confused by symmetrical D1-D2 on the schematic; they're just CR801 and 802 on the Duramodule. Also, Q4 is sawtooth shaper Q804, Q1 is Q801, Q2 is Q802, and Q803 is Q3 turned around. So here, instead of pulling a transistor or two when the oscillator fails, you'll be able to exchange an entire module at a single pull and insertion, and any oscillator fault will be repaired. Say you'd like to replace just a single transistor instead? Sorry, but that's the way the industry's going—to throwaways, pretty fast and furiously.

Let's go back to the original horizontal Duramodule (Fig. 9-23) and see how it works. Sawtooth shaper Q804 receives a negative-going pulse on the emitter that sends the transistor into conduction, discharging C803 to sharpen the initial leading

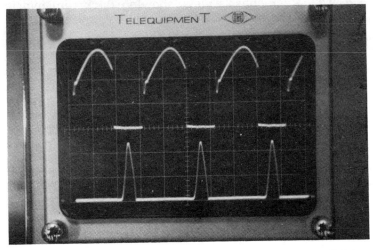

Fig. 9-21. Trace (top) and retrace waveforms for RCA's SCRs. Top trace is 100v per division, while the bottom trace is 200v per division. Time base is 20 microseconds per division.

Fig. 9-22. Rear chassis view, showing thick film hybrid circuits serving as the horizontal, vertical, and subcarrier regenerator in Zenith's 25CC55 receivers.

edge of the incoming waveform, then charging through R812 to form the remainder of the sawtooth. The negative part of this sawtooth will appear across R802, since it is coupled there by C802, but the waveform will not drop across R803 because CR802 has been turned on to place a short across R803. The reverse occurs when the positive part of the sawtooth places a short across R802 because of the conduction of CR801.

If the incoming negative sync pulse arrives at the common cathodes during crossover time, both diodes go into full conduction for an equal time and-or one is completely on and the other completely off. Therefore, the outputs will cancel, leaving nothing. However, should sync pulses lag or lead the sawtooth ramp, one diode will conduct for a shorter or longer period than the other, and the difference output will be positive or negative, depending on the difference in conduction times. C806 and R806 form an anti-hunt network to protect against overreaction at warmup.

The diode DC output makes Q801 act as a variable resistance, changing the charge potential on C809, which is in series with L214, the horizontal hold tunable inductance. Q802 is a self-starting oscillator because of forward bias coming from the 125-volt line through T7 and resistors R813 and R816. The load resistor is R818.

You may see no apparent source of collector voltage for Q801. AC sinewaves generated by the horizontal oscillator go back through All to C809. Here, these voltages are rectified by CR803 and form the supply voltage for the AFC. The oscillator output is now large enough in amplitude and sufficiently shaped that it will drive the Q803 horizontal driver completely into saturation and cutoff, so that its drive voltage, coupled through T206 to the horizontal output transistor, is the square-wave voltage that's needed for adequate off-on drive.

QUESTIONS

1. Why are AFC diode circuits so critical?
2. What is the greatest dissipation time factor in horizontal output stages?

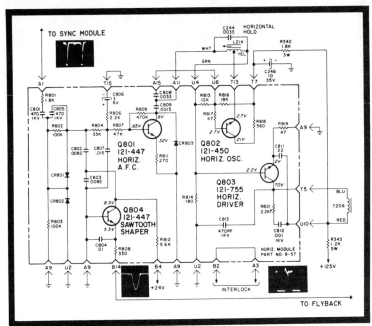

Fig. 9-23. Schematic of Zenith's original horizontal Duramodule used in the 40BC50 receiver chassis.

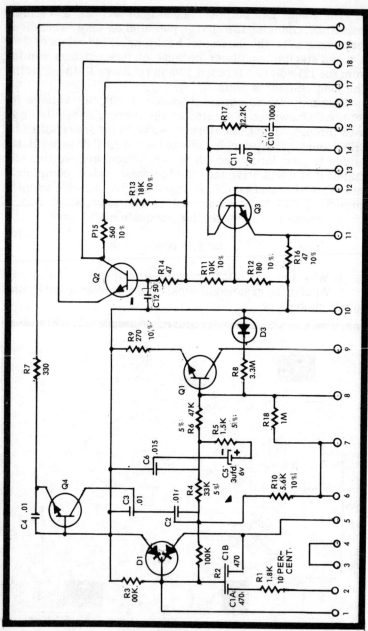

Fig. 9-24. Schematic of Zenith's thick film horizontal module that replaces many discrete portions of the Duramodule shown in Fig. 9-23.

3. What is 263 times faster than vertical sync, and why?

4. Are there any horizontal sweep systems without dampers, HV regulators, tube rectifiers, unshielded flyback transformers?

5. What are anti-hunt networks and where are they used?

6. What function has the R44-C21 time constant in Fig. 9-3?

7. The horizontal output tube cuts on at what point of the driving waveform?

8. What does a varistor do in a horizontal output tube protection system?

9. What are the three resonant circuits in an SCR drive system?

10. Are HV transformers connected in series or parallel with the deflection yoke?

11. How can an SCR be turned off?

12. Silicon controlled rectifiers are peculiar beasts. They can be turned on by a gate pulse, but how do you stop conduction?

13. In Fig. 9-21, retrace (top) and trace waveforms (bottom), are shown at 20 microseconds per div. and 200v p-p. Which SCR conducts longer and why?

Chapter 10

HV Supplies & Pincushion Circuits

The days of the color high-voltage regulator as we've known it in tube receivers has all but vanished. Its replacement is a limiter that acts not only from a high-voltage increase, but also exerts direct control on cathode ray beam current, a much more satisfactory procedure since both high voltage and beam current are simultaneously affected and receiver regulation is considerably improved. It's been common practice in the days of the 6BK4 to rotate the brightness control and watch the filament (if you could see it) of the shunt regulator to determine if it glowed as beam current increased. Even in the newer tube types, regulation wasn't that good, so as the **brightness limiter** type of circuit was developed, there has appeared a rather simple and certainly better beam current-high voltage type limiting.

With the brightness limiter came the two, three, and four section voltage doubler, tripler, and quadrupler used mainly by Magnavox in its new solid-state receiver, Zenith and Sylvania in their latest models, and RCA in the CTC46, 49, 54, etc., Series, respectively. This high-voltage arrangement has permitted smaller, lower voltage HV output transformers, less loading on the output circuits, and very adequate and predictable high voltage. Motorola, meanwhile, has stuck to the single-ended output and stacked solid-state rectifier for its transistor sets, and this operates well, too. In fact, with new horizontal output transistors being developed at a faster rate than ever, economics may well dictate a single-ended output that will bring the hefty high-voltage transformer right back again.

Also, at this writing there appears to be a considerable movement underway to develop some type of oscillator to drive all voltage supplies to provide better regulation, less filtering, possibly less space, and certainly far better efficiency at probably reduced cost. We'll look at some typical and proposed examples in the next chapter on low-voltage power supplies.

Accordingly to Jim Hornberger, Delco Electronics has developed a horizontal deflection circuit (Fig. 10-1) for a 25-inch, 110-degree color tube that supplies 25 kilovolts from a regulated 240-volt power supply. Two series-connected DTS-804 transistors are driven by a 2N3439 and the secondaries of T1 and T2. The transformer is tuned to the fifth harmonic and delivers better than 1.5 kilovolts from 0 to 1.5 milliamperes beam current. The DTS-411 insures a regulated 240-volt DC source, while the DRS-114C diode protects the transistors from excessive HV transients. The 150-meg bleeder resistor following the Varo doubler gives better low-current regulation and allows the picture tube capacitance to discharge easily when the receiver is turned off, leaving no center white spot or trace. The DRS-114D diodes are the damper (top) and clamp (bottom), respectively. A 19-inch 110-degree tube, according to Delco's Hornberger, can use only a single DTS-804 and drive the CRT at 20.5 kilovolts easily.

VACUUM TUBE HIGH-VOLTAGE GENERATORS

The horizontal sweep system (Fig. 10-2) has a number of functions it must execute, including high voltage, B-boost, dynamic convergence correction, pincushion correction, and blanking, but the most important of these is generating a linear sweep current through the horizontal deflection yoke windings. For without this sweep, there could be none of the others.

Fig. 10-3 should help you understand the operation of the output tube, damper, and deflection yoke as events occur during the line trace and retrace 63.5-microsecond interval. This illustration, you will notice, represents both voltages and currents, a none too subtle suggestion that some manufacturer should get busy and make a fairly simple, not too expensive current probe that will work with at least the better oscilloscopes, say, those with 1 meg impedances and about 30 pf shunt at the inputs. If more SCR RCA-type systems are adopted, not having a current probe will eventually lose you, believe me. There are just too many reactors, coils, and transformers to cope with.

In working with the horizontal subsystem, there are several things you should always remember: 1) retrace is caused by the magnetic field reversal in the HV-yoke coils following output tube cutoff; 2) following retrace, and for the purpose of damping the HV-yoke oscillations, the damper conducts and helps move the trace from a resting position on the left side of the screen to its center; 3) the output tube plate

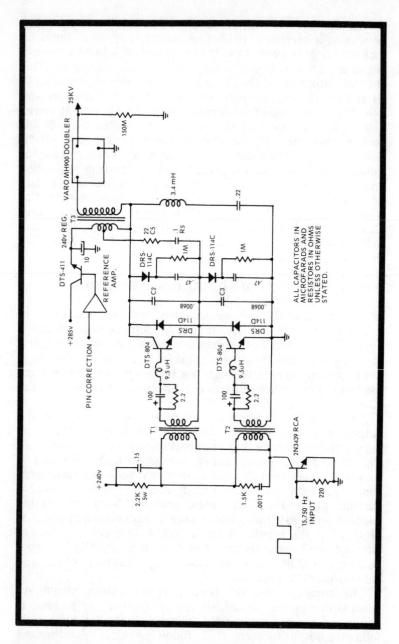

Fig. 10-1. Delco developed HV driver and output circuit for 110-degree 25-inch CRTs that will probably be seen in the U.S. between 1973 and 1974. (Courtesy Delco ElectronICS)

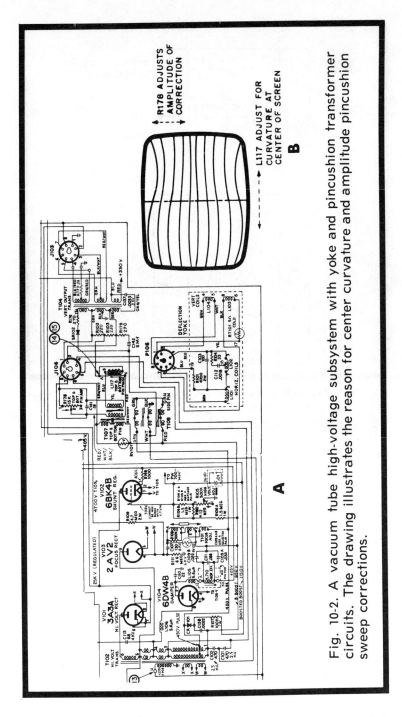

Fig. 10-2. A vacuum tube high-voltage subsystem with yoke and pincushion transformer circuits. The drawing illustrates the reason for center curvature and amplitude pincushion sweep corrections.

243

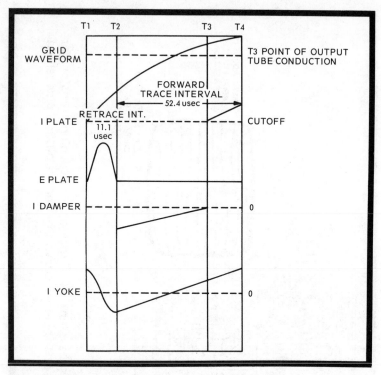

Fig. 10-3. These curves show one cycle of horizontal output beginning with retrace and ending with cutoff. (Courtesy Ray Guichard, Magnavox)

voltage rises during the time the grid goes heavily negative (T1 in Fig. 10-3), since the tube is cutoff; 4) As the **grid** (not the plate) comes out of cutoff (T2) and toward drive, the trace interval begins, the tube plate voltage is clamped, the damper current begins a linear rise, carrying the beam from center to the right edge; 5) yoke current is also linearly increasing and, when the output tube grid reaches the point of conduction (T3), the output tube plate current rises, the damper reaches 0 and cuts off, and the yoke current continues to increase linearly until time T4 and complete output tube cutoff.

Observe that the yoke current during retrace (T1 to T2) is almost a sine wave and it develops an oscillation frequency of about 70 kHz. Also, realize that even though the grid leak bias on the **grid** of the horizontal output tube begins to dissipate by time T2, it is not until the grid coupling capacitor charges to time T3 that the output tube actually begins conduction. This, then, is how all single output horizontal subsystems operate.

The SCR types are somewhat different, since they have no damper, simply trace **and** retrace complements formed through certain coils, RC time constants, SCRs, and diodes.

In the plate of the V104 damper, shown in Fig. 10-2, you will see a 5.6 microhenry coil and then a horizontal efficiency parallel circuit. The coil and its two capacitors help linearize the overlap damper-output tube currents as they pass through half-way or zero points driving the deflection yoke—the I yoke 0 point shown in the lower curve, Fig. 10-3. A check for such linearity is made at the output tube's cathode with a meter, and the efficiency coil is adjusted for minimum cathode current, often something less than 200 milliamperes.

The focus rectifier simply supplies a rectified high voltage to focus the beams in the cathode ray tube. There is a coil adjustment, since one cathode ray tube will vary from another. The various boost and boosted voltages are generated from different taps on the flyback transformer, rectified, filtered (usually) and then dispatched to their various subscriber circuits. The 850-volt boost is the reference, since any varying load on the high voltage makes the boost voltage change in the same direction.

The cathode of the regulator goes to the B+ 405-volt supply. When high voltage drops, due to beam current demand, the grid of the HV regulator becomes less positive, the regulator does not conduct as much (becomes a higher resistance) and the high voltage rises due to lighter loading. When beam currents decrease, the boost voltage rises, causing the regulator to conduct more heavily and load the high-voltage output so that HV is dropped in accordance with the magnitude of the boost supply.

Notice the arrow to PW700 MM. This extension couples video to the grid of the regulator so that when large white areas are being televised, the regulator grid sees the same polarity picture the CRT does, and the high voltage remains steady. If this coupling were not made, white scenes would immediately draw extra beam current and load the high voltage accordingly. The high-voltage adjust is simply a DC voltage divider, with the arm at ground potential.

The pincushion circuit may operate with simply a passive transformer, as shown here, or with active tube and transistor stages that also supply dynamic scan correction to the deflection coils. If it were not for this circuit, center curvature (Fig. 10-2) and amplitude beam corrections could not be made in the larger rectangular tube receivers, and the resulting image would be considerably distorted. Horizontal currents supply side pincushion correction, with vertical currents operating on the top and bottom amplitude and phase. One vertical-horizontal voltage modulates another, so that the tube extremes are affected much more than the center, and the collective sweeps are compensated accordingly.

Pincushion transformer and potentiometer adjustments can easily be made with a stable crosshatch generator. Output waveforms, as you can see, appear either as a bow tie or an ellipse, depending on sweep rate adjustment of the oscilloscope. What you should remember specifically about pincushion correction is that the yoke current is shaped to linearize the sweep rather than individual beam bending, as done by the electromagnetic convergence circuits.

ZENITH'S HIGH-VOLTAGE OUTPUT WITH VOLTAGE TRIPLER

The circuit in Fig. 10-4 is standard in the 4B25C19 (B4030-C4030) Zenith chassis series. The usual grid leak biased horizontal output tube supplies drive for the flyback, yoke, saturable reactor pincushion transformer, and AGC-AFC video-sync limiting and frequency correction circuits. The output connects to the primary of the horizontal output transformer, two windings above the V204 damper cathode tap. The yoke connects at the third and fourth tapoffs, and the pincushion reactor to the fifth. The input tube drive waveform, this time a trapezoid, is illustrated by waveform 16 (Fig. 10-4), with the AFC negative-going pulse shown by waveform 17. The most interesting waveforms, however, are those illustrated by 54 and 55.

A 500-volt positive pulse is coupled through C264 to add or subtract from the bias supplied through HV adjust R332, the 270-volt bus and grid action of the horizontal oscillator drive waveform. An increase in the amplitude of this pulse will cause varactor R333 to decrease in value, limiting positive DC to the output grid and lessening conduction. A decrease in high voltage will cause an opposite reaction, with R333 increasing in value, thus allowing the positive grid voltage to become greater since there is less divider action between R329 and R333. Consequently, the output drive becomes greater, supplying additional high voltage to make up for the drop.

The output HV inductor in this receiver is virtually an auto-transformer supplying various potentials at the tap points and B-boost through C268, one end of which is tied to the 270-volt supply, with additional current supplied by the damper for generation of the extra 500 volts, 780v being the boosted supply. The transformer, however, is smaller than the usual flyback, and its series inductance generates only about 8.33 kilovolts upon magnetic flux reversal.

The high-voltage output is coupled to the tripler circuit shown at the bottom of Fig. 10-4. This "stack" can either have an input capacitor between the flyback (usually does) and the tripler, or a capacitor to ground from terminal 8. In this arrangement, C1 charges positively through D1 on the initial HV output, but divides with D2-C2, so that both capacitors have 4 kilovolts across them. The second HV series charges C1 to 8 + 4, or 12 kilovolts, but C2 divides, and they each then store 6 kilovolts. On the third round pulse, C1 and C2 now have 8 kilovolts across them, and the second branch of the tripler begins to charge. This branch consists of capacitors C3 and C4, plus diodes D3 and D4, and they rectify and charge the two capacitors in steps just like the initial stack. Finally, the third diode rectifies and charges the capacitor, but the input potential is 16 kilovolts, since the first two stacks are 8 kilovolts each. Since C5 has no other divider capacitor with which to equalize, it retains the 16 KV plus 8 KV, and the total output applied to the picture tube is a full 24 kilovolts. Notice that the voltage is cumulative across all three stacks so that a voltmeter placed between output and ground will read the fully developed voltage, including that of the flyback transformer. The small schematics beneath the subsystem diagram show two tripler versions with slightly different divider networks and, in the first, a capacitor between the focus tap and brightness limiter.

A BRIGHTNESS LIMITER CIRCUIT

In the B-C4030 there is also a brightness limiter that controls beam current as a function of high voltage (Fig. 10-5). A beam current sampling out of the voltage tripler is connected through a potentiometer to the input of Q204, a limiter transistor, which is protected by zener diode CR214. If there is a beam current increase, the current through the arm of tripler-brightness potentiometer R338 (Fig. 10-4) will also increase and cause a positive voltage drop, turning on the brightness limiter. Q204 is a little more than a switch, and it will turn on more or less, supplying a resistive ground path for the base of second video amplifier Q205, reducing its conduction and output, and so lessening the cathode ray tube's beam current.

Positive horizontal blanking is applied through CR203 to the emitters of Q207 and Q205 which are blanked by the positive-going horizontal pulse at horizontal retrace time.

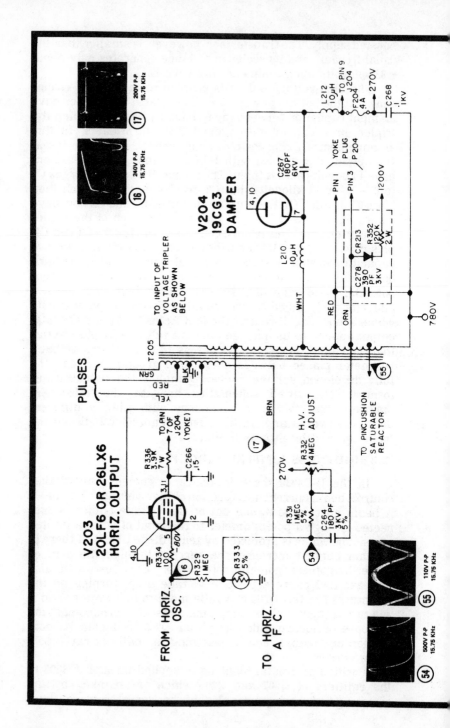

V204
19CG3
DAMPER

V203
20LF6 OR 26LX6
HORIZ. OUTPUT

PULSES

TO INPUT OF
VOLTAGE TRIPLER
AS SHOWN
BELOW

FROM HORIZ.
OSC.

TO HORIZ.
A F C

YOKE
PLUG
P 204

TO PINCUSHION
SATURABLE
REACTOR

H.V.
ADJUST

L212
10 μH

TO PIN 9
J 204

F204
4A

270V

C268
.1
1KV

C267
180PF
6KV

L210
10 μH

CR213

R352
120 K
2 W

C278
390
PF
3KV

1200V

780V

T205

R336
3.9 K
7 W

TO PIN
7 OF
J 204
(YOKE)

C266
.15

R332
4 MEG

270V

R331
1 MEG
5%

C264
180 PF
3KV
5%

R334
100

R329
1 MEG

R333
5%

PIN 1   PIN 3

GRN
RED
BLK
YEL

WHT

RED

ORN

BRN

-80V

17 ▶

16

54

55

16   240V P-P
15.75 KHz

17   200V P-P
15.75 KHz

55   110V P-P
15.75 KHz

54   500V P-P
15.75 KHz

248

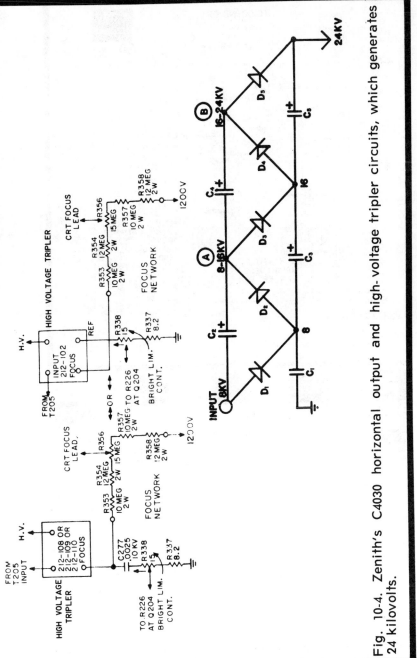

Fig. 10-4. Zenith's C4030 horizontal output and high-voltage tripler circuits, which generates 24 kilovolts.

249

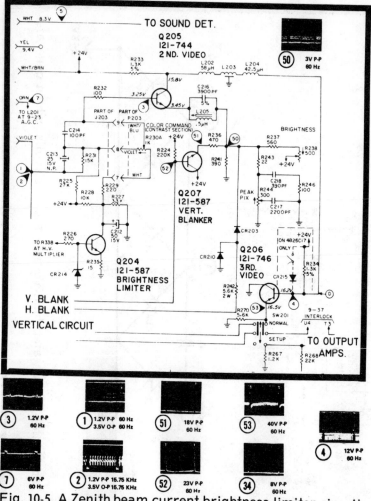

Fig. 10-5. A Zenith beam current brightness limiter circuit that controls the second video amplifier to prevent excess HV current drain.

Vertical blanking switches on the vertical blanker backbias the second video amplifier at vertical blanking time. The third video amplifier supplies the normal-setup switch that collapses the vertical sweep and shuts off video during color temperature adjustments.

PINCUSHION CIRCUIT

This Zenith pincushion circuit (Fig. 10-6) utilizes a saturable coil technique consisting of specially shaped ferrox-

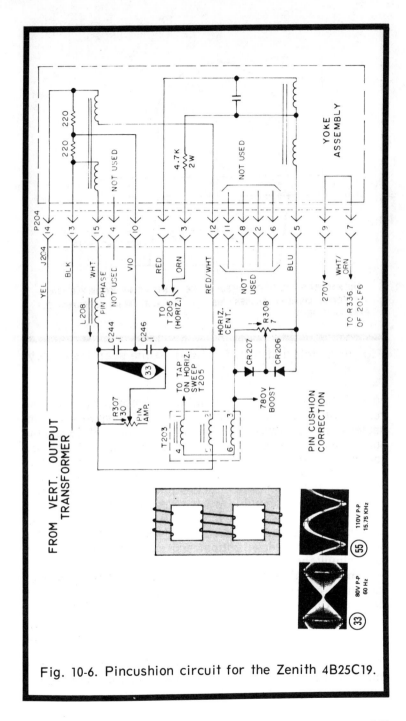

Fig. 10-6. Pincushion circuit for the Zenith 4B25C19.

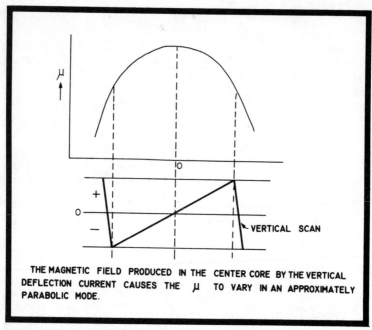

THE MAGNETIC FIELD PRODUCED IN THE CENTER CORE BY THE VERTICAL DEFLECTION CURRENT CAUSES THE μ TO VARY IN AN APPROXIMATELY PARABOLIC MODE.

Fig. 10-7. Relationship of μ (mu) to the scan frequency.

cube E-cores with windings on all three limbs (or cores). The center winding, which is used for control purposes, is connected in series between the vertical deflection coils, while the outer windings (being the controlled load impedance) are connected in series with the horizontal deflection coils and in series with each other. Connecting the two outer windings in series and in the opposite direction minimizes the interaction between the control and load windings. However, interaction remains low only as long as the core saturation is very low. At higher core saturations, interaction does occur and is used to obtain the required top-bottom and side pincushion corrections simultaneously.

The vertical deflection sawtooth current flowing through the center core winding produces a variable magnetic flux in the core. During one vertical scan, the permeability (μ) will vary from a low level at the start of the vertical scan to a high level at mid-scan and again to a low level at the end of the vertical scan. This change in permeability has a parabolic type curve representation (Fig. 10-7). The variation of permeability causes a change in the inductance value of the outer windings. Thus, the amplitude of the horizontal deflection current will vary in an approximately parabolic mode with vertical scan which will counteract pincushion distortion at the raster sides.

During one horizontal scan period, the momentary value of the vertical deflection current can be considered to be constant. Also, at the time of a horizontal scan through the center of the picture, the vertical deflection current is zero. The currents at the horizontal frequency flowing through the outer windings of the saturable reactor produce a flux in the core that passes only through the outer cores but not the center core. However, this is true only if the outer cores are equally saturated. Dependent on the vertical deflection current flowing through the center core winding being positive or negative, this current will intensify or weaken the saturation in one of the outer cores with respect to that in the other one. Due to these differences in saturation of the outer cores, the flux at the horizontal scan frequency will also partly pass through the center core.

The intensity of this flux is dependent upon the intensity of the flux at the vertical frequency (Fig. 10-8). The polarity of this flux is dependent on the polarity of the vertical deflection current. Thus, a flux at the horizontal frequency, which is modulated in amplitude and phase at the vertical frequency, passes through the center core and induces a voltage which is an approximately sawtooth shaped waveform. Since this voltage acts in series with the vertical deflection coils, it causes parabolic waveform currents at the horizontal

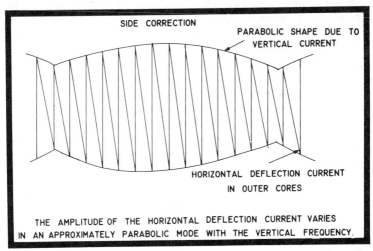

Fig. 10-8. Pincushion side correction currents used in the 4B25C19.

frequency in the vertical deflection coils. The currents have maximum amplitude at the top and bottom of the picture, but are zero at the center.

However, the polarity of this "correcting" voltage is such that the original pincushion distortion would be increased. Therefore, the polarity of the induced voltage in the center core is reversed by means of two capacitors (.1 mfd) in series between the vertical deflection coils, and the induced voltage tapped from between the coils and capacitors. Since a voltage (phase) difference exists between the voltage across a coil and one across a capacitor, the proper polarity is obtained.

The amplitude of the vertical frequency correction current (top-bottom correction) is controlled by potentiometer R307 (30 ohms) and phase controlled by the adjustable coil, L208. Proper adjustment can be achieved by setting the amplitude control for the desired amplitude for straight lines across the top and bottom of the raster and then adjusting the coil for the proper phase. In some instances, it may be desirable to work back and forth between the two adjustments a few times for optimum settings.

Horizontal centering is accomplished by applying the AC sweep current (sawtooth) through a bridge circuit in series with the deflection yoke. The bridge consists of R308, CR206 and CR207. When the arm of the control is set at its center, an equal amount of positive and negative current flows through each diode and the control. Since the current paths are equal and opposite, cancellation occurs. However, if the arm of the control is set at either end of the control, it becomes shorted by one or the other diode, depending on which direction the arm is rotated. Thus, current in one direction will be greater than in the opposite direction. Either more positive or more negative current will be permitted to flow.

HIGH-VOLTAGE BLOCK DIAGRAM

Now that we've covered some of the preceding sophisticated circuits, we are ready to tie the entire package together and show what a block diagram of the complete high-voltage system looks like (Fig. 10-7).

As you see, the output not only drives the flyback but supplies the deflection yoke, pincushion transformer, feedback to the yoke, with vertical circuit inputs to both pincushion and vertical deflection coils of the yoke itself. The high voltage may be rectified by a single tube, stacked semiconductor diode, or it may be developed from a smaller flyback transformer, then doubled, tripled, or quadrupled for a 20- to 27-

kilovolt output. The horizontal AFC is usually supplied by a winding on the flyback, while the doubler-tripler has an output to control conduction of a brightness limiter that acts on the video chain to limit beam current.

The damper (in any system besides RCA's) is placed before the flyback. Actually, the damper is in the primary of the flyback transformer on both tube and semiconductor systems and, although slightly different in circuit configuration, does the same basic job in both, except that it may or may not develop boost voltage, depending on its placement in the circuit. Where boost is generated, any (damper) diode will have to be suspended between the flyback and a high DC voltage to develop the additional high potential.

At no point, you may have observed, is there anything said about a filter capacitor for the rectified high-voltage pulse. In reality, there's a big one that amounts to about 500 pf, and it consists of the aquadag coating on the cathode ray tube. In older B&W receivers, there are still some large ceramic 20 kilovolt "barrels" to be found that need a change now and then. But in color receivers, a new HV rectifier filter means a new cathode ray tube, unless you want to innovate, and this indeed is a mite more expensive. The good old RA112-113 series of monochrome Dumont sets used to have three 470-pf 10 KV capacitors, originally oil filled, that shorted every so many years, cutting off high voltage, and producing many an interesting repair job. Capeharts used to burn up HV shunt capacitors and flyback transformers, and RCA's did their share of shorting deflection yokes and damping capacitors.

RCA'S HIGH-VOLTAGE & PINCUSHION CIRCUITS

When SCR102 is triggered (Fig. 9-9), the retrace resonant circuit is completed and the energy stored in C412, C413, and C414 causes electron flow to continue downward through the yoke, rapidly cutting off SCR101.

Resonance causes the yoke current to decay to zero and reverse, cutting off SCR102 and turning on CR102. At this instant, scan is at the center of retrace. Current increases until it reaches a maximum, driving the scan to the left side of the screen, but it cannot reverse again because gate voltage no longer is present on SCR102. The large current flowing in the yoke at the end of retrace takes the only path available, through CR101. This, of course, was the condition assumed at the beginning of this summary.

The high-voltage transformer is connected parallel to the yoke circuit. During scan time either SCR101 or CR101 is

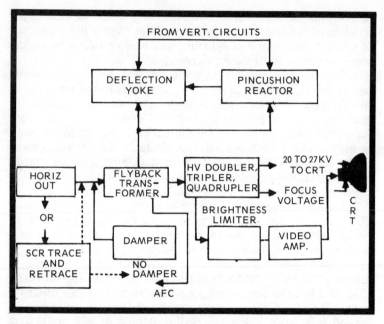

Fig. 10-9. Block diagram of a modern high-voltage sub-system.

conducting, thus shorting the primary so that very little energy can enter. However, during retrace the flyback voltage pulse drives the transformer primary, and the transformer converts this voltage to a higher voltage, which is used to drive the quadrupler and screen power supply. Energy transfer is optimized by tuning the transformer to the third harmonic of the retrace pulse. This is the function of L401.

Scan width and high-voltage level are sensed by sampling the voltage developed across C408 and C409 during scan time. An increase in scan width and high voltage increases the collector current of Q401. This further saturates T402, decreasing its inductance and increasing the resonant frequency of the **power input** resonant circuit. Since this causes the voltage at the anode of SCR102 to crest sooner and decay further before the next retrace begins, there is less energy available for scan and high voltage during the subsequent retrace and trace cycle. If scan width and high voltage tend to decrease, they are stabilized in the same way, except, of course, all processes are reversed.

To overcome side pincushion, the horizontal scan must be increased when the vertical scan is near the center of the raster. In earlier model RCA color receivers, this correction was accomplished by passive components, but because of the wider deflection angle used in the CTC49 a more sophisticated means of correction is necessary. Since the high-voltage regulator of this chassis also controls the scan width, side pincushioning may be corrected conveniently by providing a second input to the regulator. This input to the regulator is derived from the vertical deflection circuits and processed by the circuit shown in Fig. 10-10.

Q402 may be considered as a resistor which allows current flowing towards the high-voltage control to take either of two paths. One of these is through CR408 and the emitter-to-base junction of Q401, which is the normal voltage-regulating current previously described. The second path is through CR405 and the collector of Q402. Obviously, if more current flows through Q402, less flows through Q401, causing the high voltage and scan width to increase.

To correct side pincushioning, then, it is necessary only to increase the forward bias of Q402 when vertical scan is near the center of the raster and decrease it when the vertical scan is near the top and bottom. One output from the vertical deflection system is fed to the base of Q402 by way of R416 and R415 and another arrives via R417. These samples of vertical deflection signal are shaped into a parabolic waveform which reaches its maximum positive potential at vertical mid-scan.

Two inputs taken from the horizontal deflection system are used to optimize the high-voltage regulation and pin-cushion correction. The first of these is taken from terminal D (see Fig. 10-11) of the high-voltage quadrupler, via R115, and reaches the base of Q402 by way of R426 and R419. The purpose is to allow the regulator system to "measure" the beam current and regulate high voltage more accurately.

The second input is taken from terminal C of the quadrupler and reaches the base of Q402 by way of an adjusting potentiometer, R428. This input samples the high voltage by means of the capacitive voltage divider made up of C426 and the capacitors in the quadrupler, and compensates for phase shift of the side pincushion correction voltage, which is the result of the capacitance of the CRT high voltage connection.

The amount of effect which this sample from the quadrupler will have on the pincushion amplifier may be adjusted with R428. The amplifier has been designed so that when brightness is set for a barely visible raster and R428 is set to minimum (CCW), there will be no pincushion. Then, R428 is adjusted to correct the pincushioning which will appear when the brightness is increased to maximum.

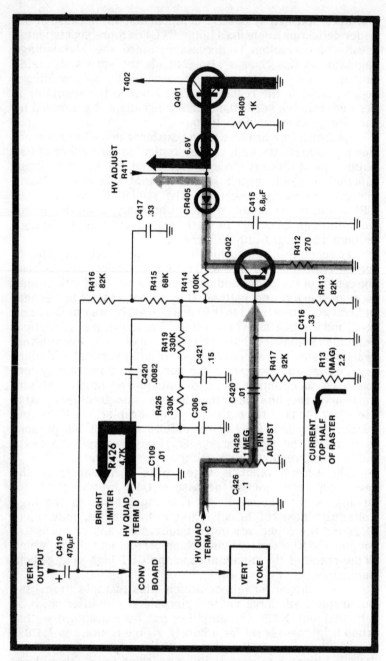

Fig. 10-10. RCA solid-state side pincushion amplifier.
(Courtesy Carl Moeller)

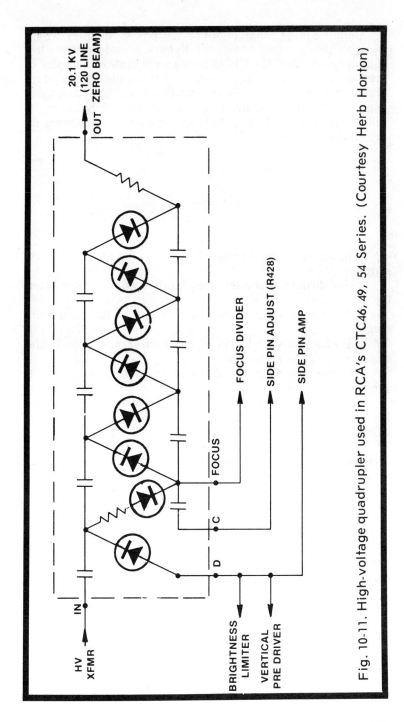

Fig. 10-11. High-voltage quadrupler used in RCA's CTC46, 49, 54 Series. (Courtesy Herb Horton)

259

Unlike conventional high-voltage power supplies which rectify a positive pulse from the flyback transformer with a half-wave rectifier, the CTC49 uses a solid-state quadrupler to produce high voltage. This reduces the required pulse amplitude from about 23KV to nominally 6KV. The quadrupler itself is hermetically sealed and is not repairable; however, its schematic is shown in Fig. 10-11 as a means of identifying its external connections.

QUESTIONS

1. What movement is now afoot regarding various power supplies?

2. Why do manufacturers use a large bleeder resistor **after** the HV output?

3. What's the most important function of the horizontal stage?

4. Do the plate and grid of the horizontal output begin conduction at the same time?

5. Varying the high voltage varies the boost voltage in the same or opposite directions?

6. When the brightness control turns on the output luminance amplifier harder, what happens to CRT beam current?

7. What happens to the high voltage when there is increased beam current? Is this the reason for brightness limiters?

8. With no picture on the screen and adequate high voltage, what do you suppose could be the next logical problem?

9. Name the newest power output drive scheme.

10. A vacuum tube type pincushion transformer, RCA type, corrects for both _____ curvature and _____ beam corrections.

11. What's so important to remember about a pincushion transformer?

12. In Fig. 10-4, use the equation:

$$E_0 = \frac{E\ in\ x\ R\ 333}{R329 + R\ 333}$$

and solve for R333, if Eo is 10v, Ein is 200v and R329 is 1 megohm. Try another value for good measure.

13. If R333 was 1K and you wanted an output of 12 volts, what would have to be your input potential?

14. Brightness limiters do what?

15. Vertical sawtooth current through the core of a pincushion transformer produces a _____ _____ At scan, the permeability will vary from low on each end to high in the center and vary the horizontal deflection current.

16. What happened to the damper in RCA's newest solid-state receivers?

17. Why did RCA have to go to an active pincushion circuit in its CTC49 chassis?

Chapter 11

Low-Voltage Power Supplies

Low-voltage power supplies used to be, in the days of simple vacuum tubes, just a matter of whether you had a "hot" or "cold" chassis and, perhaps, whether rectifiers were either half-wave or full-wave. In the less expensive receivers, this outlook hasn't changed much, because of the lower cost involved. But in the better color sets, there's now a considerable choice with many, many engineering advances in the past two years and, apparently, considerably more to come. For instance, the industry is talking right now about switched power supplies for everything but the collector voltage of the horizontal oscillator, and this could be the only directly rectified feed. All other voltages could come from power circuits that would be driven, probably, by the horizontal oscillator, or higher or lower frequencies thereof. A big advantage would be better short circuit protection, which is very important with semiconductors. Others include much less filtering at the higher frequencies, a completely modularized power supply, less costly low-voltage transformers, and reduced ripple on the bus feeders, to name a few. Motorola, in fact, is already doing just this, and has additional ideas in the engineering think tank for the future. RCA, also has a partially plug-in power supply, with regulators among the modules, and the removal of modules in certain Zenith receivers will disable DC immediately.

Romancing power supplies has been a pretty dull job until recently, but the already tremendous setup and competition in consumer products will certainly result in many unique and useful innovations now and in the years ahead. And it may just come to pass that the active filter principle, used in the past several years by Sylvania, may be well worth adding to the very advantageous oscillator switching method of developing power.

POWER SUPPLY BLOCK DIAGRAM

The block diagram in Fig. 11-1A is typical of the power supply found in better receivers. AC passes through the line cord into the receiver interlock and coil line filters, then through the voltage-dependent resistor that, when cold, exhibits little resistance. In a second or two, the screen area is degaussed, but no substantial AC reaches the isolating stepup or stepdown transformer. In a moment, however, the thermistor gets hot, decreases quickly in resistance, and current passes directly to the transformer primary while the VDR blocks all AC from the degaussing coils. The transformer secondary supplies the rectifiers, which usually pass the positive alternations and block the negative ones. The resulting DC ripple is now filtered by chokes, resistors, and capacitors and passed, non-regulated, to circuits like the audio output, CRT, etc. The more critical portions of the receiver, such as transistorized IFs, video amplifiers, and the chroma circuits—especially if these are semiconductors—receive regulated supply voltages.

In the less expensive version in Fig. 11-1B, we didn't bother to put in a degaussing coil, since it would be the same principle, but we do show a half-wave voltage doubler, with one side of the AC line connected directly to receiver chassis. This is the "touchier" type power supply where—if the AC plug has the right (or the wrong!) polarity and you have your hand on ground or a cold water pipe—you could become an instant angel with no difficulty. **ALWAYS use an isolation transformer** when you're working with one of these, not only to protect yourself but your test equipment, too. Well-regulated changes of 20 or 30 amperes and 115 to 120-volts (some 2300 to 3600 volt-amperes) play few favorites. Most unfortunates never live to recognize their mistake.

In case you meet this voltage doubler in your work, here's how it works: On the negative alternation, C1 charges through D1 to 150 volts. On the positive alternation, C1 discharges through D2, charging C3 and C4 by way of L1 (for C4). The 120-volt input, naturally, is doubled through a 150-mfd capacitor, and comes to D2 at about 300 volts. Then there's a 20- to 40-volt drop because of rectification loads, and dividers. The capacitors are low impedances to AC ripple, while the choke is a low-impedance to DC and a high impedance to AC. The ripple frequency, in this instance, is 60 Hz, always a tough one to filter, while ripple in the supply diagrammed in Fig. 11-1A is 120 Hz, since you'd have full-wave or bridge rectification. The faster the frequency the less and more uniform the ripple and the easier it is to smooth out by filters, where component values decrease somewhat in proportion to frequency. Remember;

$$C = 1/2\pi f X_c$$

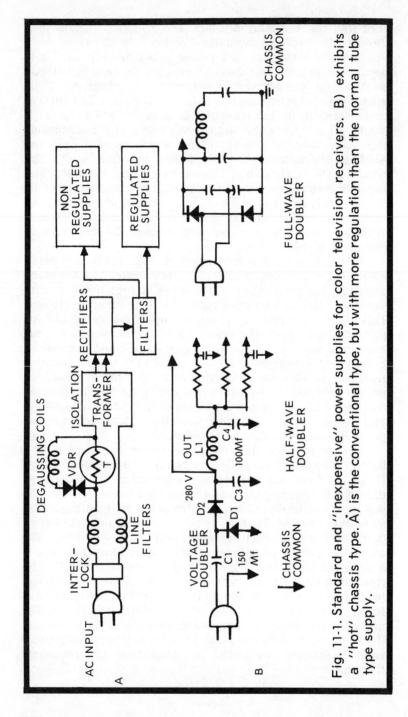

Fig. 11-1. Standard and "inexpensive" power supplies for color television receivers. B) exhibits a "hot" chassis type. A) is the conventional type, but with more regulation than the normal tube type supply.

264

You might also recall that RMS is the equivalent sine-wave energy for DC power; then RMS X 1.414 equals peak, and RMS X 1.414 X 2 equals peak-to-peak.

SEMICONDUCTOR BRIDGE POWER SUPPLY FOR VACUUM TUBES

Since solid-state diodes generate much less heat, are relatively reliable, and maintain their efficiency to the point of life expiration, the more recent color receiver tube power supplies employ this type of rectifier. Also, a bridge rectifier is more effective and with it, a smaller power transformer can be used instead of the very heavy type associated with the usual full-wave diode or tube supply.

In Fig. 11-2, we have a diode bridge rectifier that receives AC from the secondary of the power transformer after current has passed through RF chokes L115A and B and the associated RC filter networks on either side of the input line. Normal voltage input goes to the BLK-RED tap. In high line-voltage situations, the BLK-WHT tap is used.

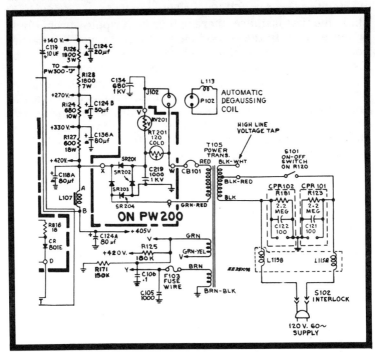

Fig. 11-2. Standard unregulated bridge-type supply circuit for vacuum tube receivers.

Current passes initially through the degaussing coil until thermistor RT201 heats and shunts AC directly to SR201. On a positive alternation at the top of the transformer winding SR201 conducts, SR202 and SR203 are back biased, and the capacitors in the voltage divider string and on the other side of choke L107 charge. Current returns to the lower half of the transformer secondary through ground and SR204. When the lower (GRN-RED) half of the transformer is positive, SR203 conducts, SR201 and SR204 are back biased, and the current return to the upper half of the transformer is through ground and SR202. All capacitors are fully charged after several alternations, degaussing, of course, is cut off, and the receiver "sees" all voltages, including the +420 volts of virtually unfiltered supply that furnishes DC potential for the cathode ray tube filaments. The 420-volt supply keeps the CRT cathode-heater potential close enough to prevent serious arcs and consequent damage to the tube.

In troubleshooting such a supply, **first** measure the various voltages, **then** look at the AC ripple wherever a voltage is down. When a capacitor fails to charge, is leaky, or shorted to ground, you'll have a lower voltage or no voltage at all. In the first instance, there will be high ripple—probably 10 or more volts. In the second case, obviously, you'll measure nothing or almost nothing, and the capacitor will have to be disconnected to prove its condition. However, when a capacitor shorts, it usually draws current through a resistor which burns, or at least smokes. **Don't** just replace the resistor and pray; measure resistance along the line where the burned resistor is connected and discover why. And **do not** use an ohmmeter to measure capacitance, either for opens or leakage; it simply can't be done. Use a capacitance checker or a current meter, but **never** an ohmmeter! An oscilloscope with an LC probe will show you any measurement of AC ripple. And any excessive ripple can always be identified by **shunting** a filter capacitor across the line to ground. If the ripple subsides, you can go find the bad capacitor. If it doesn't, check the frequency of the ripple and find out what section of the receiver is radiating. **Then** find your bad filter and fix it!

When—and this applies strongly to semiconductor power supplies—you have lowered voltages along one or more power feeder lines, disconnect each line, one by one, until the load returns to normal. This identifies the problem circuit. Next, you have to find the stage, and then the cause of the trouble in that stage.

In power supply troubleshooting, **the method you use is king**. These solid-state receivers cannot be shotgunned. If you don't believe me, try it. Your shop will pay a big bill to have your work done over, and you may not have a job. We deliberately discussed troubleshooting here to make you very aware of the basic methods for handling troubles in power supplies. In some of the more sophisticated DC sources we'll be talking about, you'll have to go far beyond the basics to handle their special problems efficiently. The "disconnect" procedure, however, is always a successful technique in any complex that is either a high supplier or high user of current—and **never forget that a transistor is a current-operated device** with DC biasing its most sensitive and demanding parameter. The reason is that a bipolar semiconductor is substantially a low-impedance device, while a vacuum tube, and its semiconductor cousin, the FET, are voltage-operate high-impedance devices, at least input-wise.

It is worth remembering that in a bipolar: Ie (emitter current) equals Ic (collector current) + Ib (base current) and that the static current gain equations for the three transistor configurations are:

$$h_{fb} = \frac{Ic}{Ie} \qquad \text{common base}$$

$$h_{fe} = \frac{Ic}{Ib} \qquad \text{common emitter}$$

$$h_{fc} = \frac{Ie}{Ib} \qquad \text{common collector}$$

while the common-base current gain is:

$$Alpha = \frac{Beta}{Beta + 1}$$

common emitter current gain is:

$$Beta = \frac{Alpha}{1-Alpha}$$

and

$$B + 1 = \frac{1}{1 - Alpha}$$

Therefore, Ic equals BIb, or in the full expression:

$$Ic = BIb + (B + 1)(Icbo)$$

taking into account the leakage from collector to base with the emitter open. In a vacuum tube, amplification factor (mu) equals transconductance (gm) x plate resistance (rp); so:

$$gm = \frac{\mu}{rp}$$

and

$$\text{Gain } A = \frac{\mu ZL}{rp + ZL}$$

In a field-effect transistor; gm equals Id (drain current) divided by Vgs (voltage, gate-to-source) and Av (voltage gain) equals gm x ZL.

In all this, you can begin to see the similarity between FETs and tubes, except the FET, and especially the MOSFET, a metal oxide semiconductor, are even more useful devices than the vacuum tube. They can do more things, with great bandpass and speed, and still don't need the huge voltages and currents most vacuum tubes must have.

ZENITH'S 25CC55 REGULATED SOLID-STATE SUPPLY

This 25CC55 power supply offers a separate winding on the secondary for the degaussing coils, extra filtering in the regulator portion, and a new limit switch. The limit switch is interesting in that high-voltage pulses from the flyback are rectified by CR210, then filtered by C242. The remaining DC, when it becomes large enough, fires the 3AGB PL1 lamp, and this causes Q209 to conduct to ground, shutting off the regulated 24-volt supply for as long as PL1 is lighted. This now makes the receiver comply with FCC regulations. Otherwise, under normal operation, Q209 is not in conduction.

With Q209 off and Q212 producing normal regulation, R324 can be considered the load resistor. R326 is the source-to-base bias, load and limiter for reference zener CR213. C241 is a filter. Changes in the load or collector voltage are referenced to CR213 and the transistor corrects its voltage output accordingly. The protection resistor is 20-watt R322, which will take most of the current if there is a short through either Q212 or Q209, with F203 as a final fuse backup in case of extraordinary overload.

The voltage-sensing circuit also has a bit of a turn-around compared with earlier supplies, with B+ adjust in the emitter of Q214, a thermistor between emitter and base, zener reference in the base, and a 250-volt voltage input bus that is rectified and filtered from a source in the primary of the flyback transformer. CR222 through CR225, the main bridge rectifiers, do exactly what the previous bridge did, this time with CR223 and CR224 conducting first, then CR225 and CR222. Return by way of ground passes through interlocks on the chroma, 3.58-MHz subcarrier, AGC, and vertical mdoules, so that any break in this continuity shuts off DC for the entire power supply. The horizontal oscillator is on another DC hot line at the output of the bridge, and a defect in those circuits will cut off the voltage driver, regulator, etc.

Voltage developed at the output of the bridge is 160 volts, which is filtered to supply the audio stages and the voltage driver and regulator transistors. As is shown in Fig. 11-4, the side pincushion correction circuit is a vital part of the regulated power supply. A 60-Hz parabolic waveform is supplied by the vertical output transformer, but with a large positive spike that is blocked at the base of Q211 by CR212. The pure parabolic voltage is now amplified and inverted by Q211 and is AC coupled to the base of voltage sensor Q214 (Fig. 11-3). This transistor's base reference is zener diode CR215, with bias supplied from both emitter and collector by a network of resistors that also supply a DC component from the flyback 250-volt source. The DC output is set by R338, the B+ adjust. This feeds the base of Q213, which, through its emitter, supplies the base of regulator Q215.

Since the reference diode has a fixed voltage (in one case, 140 volts and in the other 24-volts), any change in voltage will

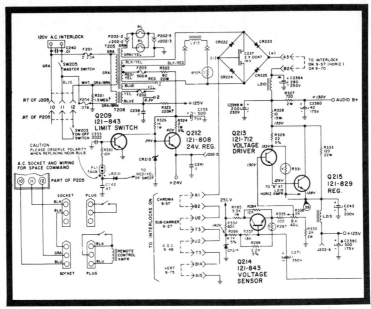

Fig. 11-3. Revised power supply appearing in Zenith's all solid-state 25CC55 receivers. PL1 lamp stays lit with overvoltage HV surge until the receiver is turned off, then on. PL1 when on kills the +24-volt supply.

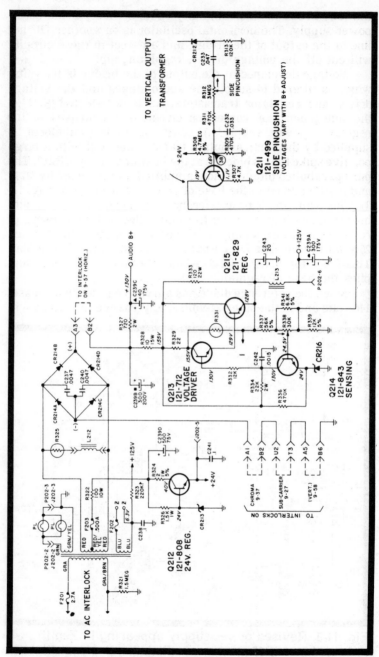

Fig. 11-4. The original low-voltage supply circuit for the 40BC50 Zenith and side pincushion correction circuit.

result in a change of current through Q214, reducing or raising the bias on Q213, and so regulating the output of Q215. Below the AC protection varistor, you will see an arrow to the horizontal output transformer and you will also observe the side pincushion amplifier bus coming into the circuit between R337 and R338. What happens here is that the vertical 60-Hz parabola actually modulates the B+ to the horizontal output transformer and expands the B+ with vertical scan to change the picture width to compensate for pincushion distortion by pushing out the sides of the picture through its middle at a 60-Hz rate. Filtering for the resulting 60-Hz induced ripple is taken care of by R327, C239C, and L213, and C239A and B.

RCA'S CTC46 SUPPLY

The generally sophisticated circuit in Fig. 11-5 has the usual interference reject coils, circuitbreaker and LC filters, high-low line switch, and isolation line transformer with five voltage sources that are all rather carefully filtered and rectified. The individual video and chroma plug-in modules have their separate or collective regulators, instead of single source regulation from the power supply. Both diode bridges work as all bridges do, and there is a thermistor at terminals 4 and 5 to block and shunt current when cold and hot, respectively. After degaussing has been completed, the half-wave CR1 is allowed full current and the 225-volt source comes on to provide drive for the set. A centertap off the 77-volt supply produces the 26- and 30-volt sources for the sound and other circuits.

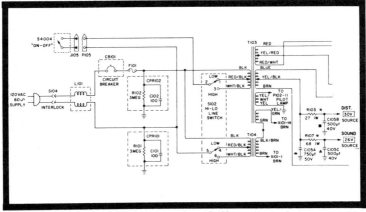

Fig. 11-5. RCA's CTC46 supply. Voltage regulators are on various module plug boards.

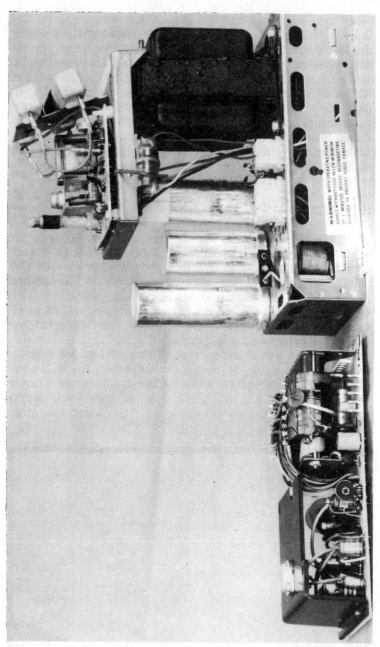

Fig. 11-6. Comparison between Motorola's new switched plugboard power supply and much larger AC-regulated unit used in the Original Quasars. (Courtesy Motorola)

One of the handy features of this supply, however, is the plug board (part of PW600) that contains all rectifiers and the large thermistor for the degaussing coil. If you suspect any of these elements being faulty, simply pull the MAB module, substitute another, and find out.

MOTOROLA'S SWITCHING POWER SUPPLY FOR CTV7-8 CHASSIS

Like the "offshore" beginning of Instamatic, Accu-Matic, or by whatever name present chroma and other controls have become known, this general power supply had some foreign origins, too, but its U.S. introduction is very likely the beginning of an era of all sorts of switching power supplies. Why? Because of utility, less expense, lighter power transformers, stepdown instead of stepup voltages, and the other advantages named earlier in this chapter. To this Motorola power supply they all apply. And, since a picture can sometimes tell more than 1,000 words, look at the difference in Fig. 11-6 between Motorola's new and original all-AC regulated Quasar power supplies in terms of filtering, size of components such as capacitors, the power transformer, huge dropping resistors on top the regulator subchassis. In addition, almost the complete supply is made of plug-in modules and can be removed by simply unplugging panel JA in any of the Quasar CTV7 and CTV8 solid-state receivers.

The schematic for the switch mode supply is shown in Fig. 11-7. As you can see, except for full-wave recitifer diodes, filament transformer, several other coils, and the degaussing units, most of the circuitry fits snugly on plug panel JA.

The input voltage doubler is made up of D800, D801, and capacitors C800 and C801, collectively, that are attached to one side of the AC line. C800 also goes to isolated ground. Rectified AC charges C800A, C801 alternately plus the other capacitors, and D800 passes a total WVDC about 300 volts that, because of drops and loads, becomes 280 usable volts. This filtered DC is now brought over the top of panel JA, in through terminal 10, to the primary of low-voltage transformer T3. (In block diagram form, this DC connection and the ensuing circuit description is shown in Fig. 11-8. You may follow both the schematic and block outline together.) DC energizes the primary of the transformer through the action of output switch Q8 to isolated ground. This switch is driven through a series of three other transistors plus a regulator, and all are protected by a silicon controlled rectifier (SCR) crowbar microsecond shut off.

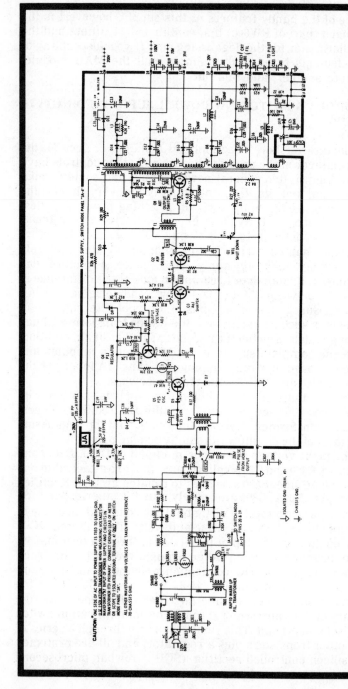

Fig. 11-7. Motorola's switch mode power supply schematic diagram. (Courtesy El Mueller)

A 200-volt sync pulse from the 15,734 Hz horizontal output appears across the primary of T2, a blocking oscillator, connected between collector and base of Q5. Thus, the oscillator is synchronized with the receiver's horizontal oscillator and its output is shaped as a rectangular wave to control the driver and, finally, the output switching transistor. The action is further shown in Fig. 11-9, along with the various waveforms that may be expected. These waveforms are, of course, exhibited for normal conditions only when regulation is unnecessary. But when, say, brightness is increased and the secondary of the transformer needs more power, the Q8 DC switch is closed for a longer time period and primary current increases, thereby satisfying the secondary demand.

The switching transistor is controlled as follows: Regulator Q4 takes samples of the driven oscillator voltage from a separate winding (pins 3 and 12) on T3, which is routed back to the base of the regulator through diode D15. At the same time, positive DC from the 33-volt bus, regulated by zener D4, and filtered by 50-mfd C18, is the reference DC for the base and emitter of Q4. Another reference is series RC circuit C24 and R35, which is connected between the Q4 base and the 280-volt line, establishing a positive-to-negative charge on the capacitor that can vary with DC voltage changes. Further RC time constants C12, R30, R16, etc., help shape this feedback voltage in the Q4 base-emitter circuit. Any

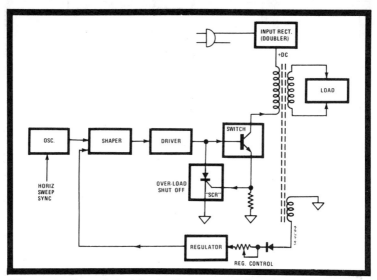

Fig. 11-8. Complete block diagram of the electronic power supply. (Courtesy Chuck Preston)

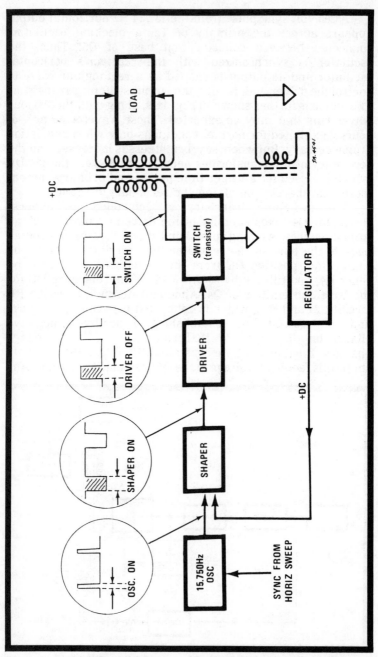

Fig. 11-9. Switched power supply operational diagram with rectangular waveforms. (Courtesy Stan King)

changes in the DC supply will produce an RC change at the base of Q4, resulting in a steepening or lengthening of the sawtooth at the emitter of the regulator, changing the width of the oscillator output waveform and so shaping the final drive pulse to the output switching transistor.

Under extra load conditions, the output switch is kept on longer, therefore, supplying more current to both the primary and secondary of the transformer so that the response promptly fulfills further current drain requirements. Conversely, less supply needs cause the regulator to conduct for shorter periods and so narrows the rectangular drive waveforms for the output switch, keeping it on for a shorter period and so reducing current in the supply.

Pretty much the same situation occurs if the line voltage increases or decreases, and so the supply is regulated both from within and without, and any internal supply transients on the various buses will appear stationary at a 63.5-microsecond time rate. Should the voltage through the output switch reach a level that causes D3 to zener and pass a signal, this will turn on SCR Q1 and the entire supply will shut down until the receiver is turned off (or the circuitbreaker blows) and the anode of the SCR has the voltage removed entirely. Gate voltage turn on will never turn off an SCR, only the removal of its anode potential.

DELCO'S FLYBACK SWITCHING REGULATOR

To show that other companies besides Motorola are already actively thinking about a switched power supply, consider the following material, developed by W.A. Robinson of Delco Electronics. Instead of a blocking oscillator, however, a self-generated Colpitts (two capacitors) oscillator (Q1) is used to produce a sine wave. The Q1 and Q2 emitters and the collector of Q3 are regulated by zener diode D5. Feedback modulation voltage comes from winding "LT" directly into the emitter of Q2. The feedback voltage and the sine wave are mixed between R13 and R14. This feedback is proportional to the DC output voltage. There is also a feedback effect through R2 in terms of supply voltage, and R2 is also the starting drop resistor for the Colpitts oscillator. Voltage adjustment R11 is located between R10 and R12. R13 actually determines the maximum pulse width, which goes to Q3 (a shaper), to change the sine wave into a pulse of varying width, depending on the feedback and DC reference modulation. Thereafter, driver Q4 supplies energy for the primary of T1,

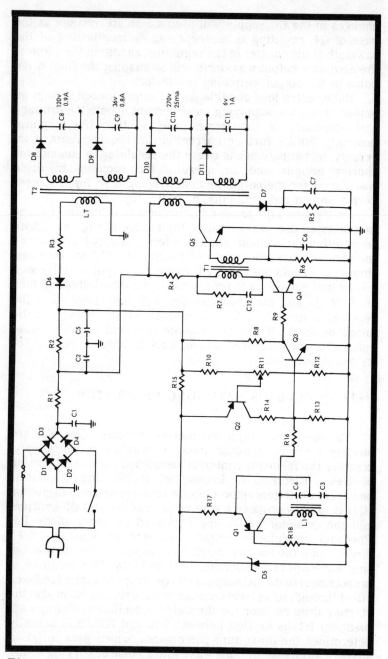

Fig. 11-10. Delco's flyback switching regulator circuit for
large screen color receivers.

which is flux coupled to the base of switch transistor Q5, the primary switching source for T2. Components R5, D7, and C7, a voltage clamp for Q5, absorb the higher than normal voltage spikes.

Diodes D8 through D11 supply the complete current and operating voltage outputs as shown on the schematic: 120 volts at 900 milliamperes, 36 volts at 800 milliamperes, 270 volts at 25 milliamperes and 6 volts at 1 ampere. The entire circuit operates at 20 kHz with small T1 and T2 transformers; the first is a laminated steel core type and the second has a ferrite core.

QUESTIONS

1. What is a "cold" chassis, and why do the less expensive receivers have one side of the chassis connected to the AC line?

2. How does C1 in Fig. 11-1B charge on the negative current alternation?

3. A sine-wave peak-to-peak voltage of 340 volts equals what value at RMS?

4. Why are so many 4-bridge diodes used in modern power supplies?

5. What happens when a degaussing thermistor gets hot?

6. When a capacitor in a low-voltage supply opens, what problems occur?

7. Why do you always use an ohmmeter to measure coils and capacitance?

8. The collector current is the sum of the ———— and currents?

9. Show that transconductance equations for FETs and tubes are somewhat similar.

10. Why do you suppose Zenith added a diode-transistor limit switch in the power supply?

11. How does Zenith take care of its solid-state pincushion problem?

12. RCA differs in its power supply philosophy from other set manufacturers. How?

13. Is Motorola's new power supply DC or AC governed in it's regulation function?

14. How do the Delco and Motorola regulators basically differ?

Chapter 12

Chroma Circuits

So much has been written and re-written about the chroma section in color television receivers that we felt compelled to avoid complicated, involved explanations in favor of a very simple, straightforward presentation that should stand the test of time. And after you have explored the block diagram, diode, transistor, and then integrated circuit demodulation, along with general color processing, you should have a very thorough idea of most, if not all the methods in use today and the many reasons for their existence. We begin with a block diagram, and then examine each section individually.

THE CHROMA SUBSYSTEM

There are three inputs to any U.S. chroma subsystem (Fig. 12-1). No. 1 is the chroma input from the first or second video amplifier. No. 2 is the burst, which comes from either the first bandpass amplifier or directly from the video amplifier, depending on the amplification needed. And No. 3 is the high-voltage transformer-generated gating pulse that turns the burst amplifier on.

Following the burst amplifier, there are often two comparators, each usually composed of a pair of diodes. They match the phase of the incoming color sync burst against feedback from the 3.579,545-MHz subcarrier oscillator, and correct the oscillator frequency. Also, there is a second comparator, doing the same thing as the first comparator, but this one generates a DC automatic control voltage, reputedly a linear function of incoming burst amplitude, that is supposed to proportionately regulate the amplitude swing of the bandpass amplifier(s). The bandpass circuits, naturally, have a chroma amplitude control, while others also have a tint or hue control.

The burst amplifier is gated into conduction by a comparatively large pulse from the flyback transformer every 63.5 microseconds for a period of between 5 and 10 microseconds to admit the color sync burst. This burst amounts to 8 to 11 cycles on the back porch of the horizontal

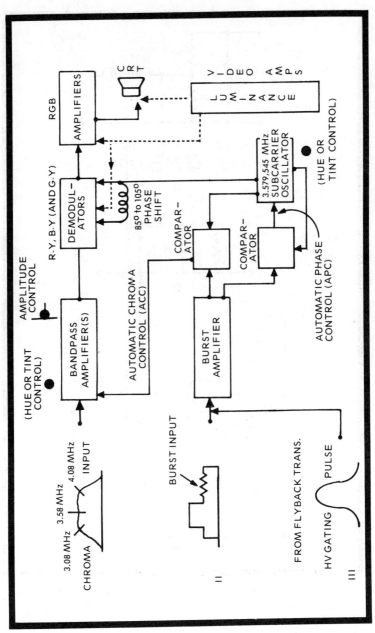

Fig. 12-1. Block diagram of the chroma subsystem found in any U.S. made color receiver. The arrows point in the direction of signal flow.

281

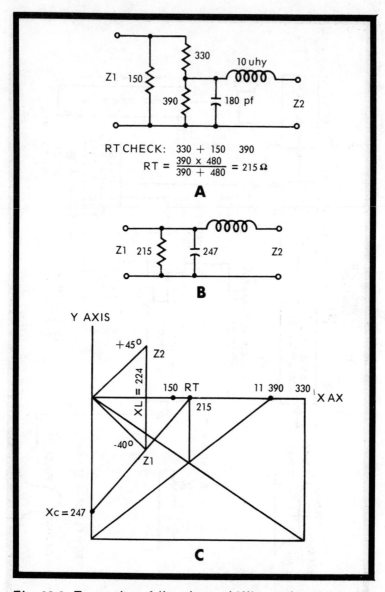

Fig. 12-2. Examples of the phase-shifting network used to divide the phase of the subcarrier oscillator inputs to the R-Y and B-Y demodulators. Network B) is the equivalent of A), and C) is the graphic analysis of either. Vertical line symbols before 390 ohms mean parallel. The resulting phase angle difference between Z1 and Z2 is 85 degrees.

sync pulse and on top of the blanking pedestal. There may be a hue or tint control that manually governs the phase of the 3,579,545-MHz subcarrier, rather than the one bracketed in the bandpass amplifier; this is a choice among the various manufacturers, but usually the phase of the subcarrier is shifted since it's a simple sine wave and easier to control.

The bandpass information now goes to the demodulators. Also applied to the demodulators are two subcarrier oscillator signals, one at 85 degrees and the other at 105 degrees from burst. These phase angle separations are, however, entirely predictable, either by straight math or through a relatively simple system of graphic circuit analysis that we'll demonstrate. The circuit, along with its analysis, are both shown in Fig. 12-2. Initially, we see a 180-pf capacitor in parallel with a 390-ohm resistor. The 390-ohm and 330-ohm resistors (in series) are paralleled with a 150-ohm resistor. Finally, there's the 10 microhenry series coil in the Z2 output.

First, let's show the sum of the 150- and 330-ohm resistors on the graph—no phase shift. We'll drop a perpendicular from the 330-ohm resistor to some convenient point, cross over to the Y axis, then draw a line from 0 X-Y to the right corner of the diagram and another line from the parallel (‖) 390-ohm resistor to the left lower corner in the third intersection. Where the three junctions meet, and directly up to the X axis, is the resistive (RT) total. Now, we calculate the reactance of the 180-pf capacitor: $X_C = 1/2\pi fc$

and this amounts to 247 ohms, which is promptly measured on the 4th quadrant negative Y axis. A line is drawn between this point and RT, and a perpendicular laid off to the 0 X-Y junction on the left. This angle, measured by protractor, is -40 degrees. The inductive reactance:

$$X_L = 2\pi fL$$

is calculated at 224 ohms, and, because it is inductive, is laid off vertically according to the measured distance, and a diagonal is again drawn to the 0 X-Y axis. The measured angle here is 45 degrees. So, from -40 to +45 is an algebraic total of 85 degrees, and this is the same type of phase angle shift we show accomplished by a simple coil in Fig. 12-1.

With an adequate vectorscope, you would see this phase angle precisely reflected in the vector output as the phase difference between the third R-Y and sixth B-Y vector petals (Fig. 12-3). At 85 degrees, this angle would be less than quadrature, while at 90 to 105 degrees it would be equal to or greater than quadrature. These phase angles are determined at times by some of the "company official viewers," while at

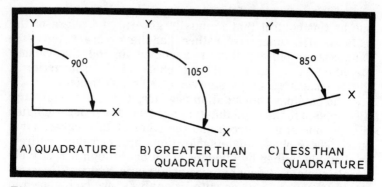

Fig. 12-3. Angles of demodulation used in the past and present to drive R-Y, B-Y and sometimes G-Y, although G-Y is always a function of the reds and blues. A) quadrature was Zenith's original; B) now the phase angle used by almost everybody; C) RCA's most copied X and Z original.

other times—and this is the more prevalent condition now, thank goodness—by the design engineers who are trying to find the best angle to complement the latest RGB phosphors. With yttrium phosphor reds, the usual angle of demodulation is very nearly 105 degrees. Green now, however, is the weakest phosphor, and so the newest receivers "key" on this color instead of red, as most did in the older tubes when blue and green were the stronger phosphors.

The "demodulator evolution" began with RCA's X and Z type, which demodulated at an angle less than 90 degrees and used relatively few components. Some off shore and USA manufacturers continue their use even today. Next were Zenith's original high-level demodulators, which amplified first (the opposite of the RCA system) then demodulated, passing the demodulated chroma directly to the electrodes of the picture tubes. Both these systems, by the way, fed the color signals to the grids of the picture tube (Fig. 12-4) and luminance into the cathodes; the picture tube actually did the RGB and Y matrixing (mixing). Both these systems derive the greens from negative portions of the R-Y and B-Y inputs (Fig. 12-5A through E). The second innovation (B) used principally by Motorola and RCA, was the 3-diode matrix, where G-Y was actually demodulated separately, but RCA still went to the cathodes and grids, while Motorola put pure RGB directly into the CRT cathodes and grounded the grids at the AC level.

Next, Motorola (C) put chroma and luminance together into an integrated circuit, and went into the CRT with pure RGB. At about the same time, Zenith (D) began demodulating R-Y and B-Y with an IC, matrixing the Y and greens, then amplifying before application to the cathodes of the CRT. Finally, in 1970, RCA and Zenith "IC'd" (E) the entire chroma and subcarrier color sections, and fired the CRT cathodes directly with reds, blues, and greens.

The reason for going to cathode CRT operation, according to Fairchild's Normal Doyle, is that with the tube matrix system, gray-scale tracking adjustments are made in the luminance channel. And this requires qualification in the demodulation process because of the difference in phosphor efficiencies and to relate the grid drive more nearly to the cathode drive. In the red, blue, green system (RGB), however, color-difference signals are mixed before entering the CRT, and only the cathodes are driven, while the grids are at AC ground. In this method, both signals are already arriving at the same set of tube electrodes simultaneously; therefore, no individual correction is needed for best color reproduction.

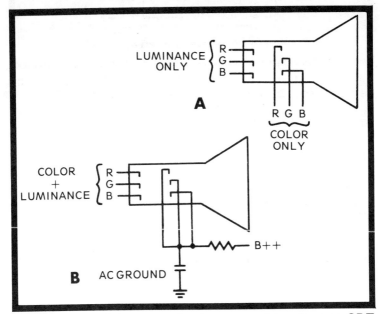

Fig. 12-4. Old style separate chroma and luminance CRT inputs (A) to the grids and cathodes. In the newer version, all drive signals go to cathodes of the CRT, with control grids at AC ground (B).

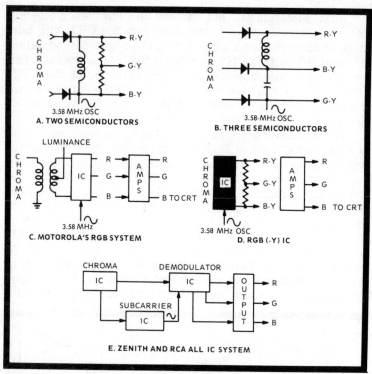

Fig. 12-5. Chroma demodulators in their order of development between 1955 and 1972. C and D were developed simultaneously. Actually, there's not an enormous difference between the two latter systems.

MODULATION TO DEMODULATION

Although we talked about this subject somewhat in Chapter 1, more detail is needed so you will be able to translate between the two with ease and certainty.

Since we cannot transmit two separately modulated subcarriers of different frequencies because of non-frequency interlace and the resulting beat that would ensue, we resort to two-phase modulation of a single carrier. These "two" subcarriers are of identical 3,579,545-Hz frequencies but are separated in phase by 90 degrees. In reality, just as in the receiver, this is actually a single carrier generated by a single source, with its slave branch phase-shifted 90 degrees. Fidelity of color transmission and reception can be greatly affected by crosstalk among all the primary colors if the transmitted phase separation is not carefully maintained, and if the receiver is poorly designed or has component defects. If, for instance, there is carrier imbalance, a gray or white picture area will not cause the transmitted I and Q signals to go to zero, and these neutral shades become colored and are quite objectionable. Also, if there is a video unbalance, undesirable video will be added to luminance and distort the gray scale. A positive upset in the Q modulator, for example, would somewhat intensify reds and blues but darken greens. Negative unbalances would intensify greens but shade blues and reds.

Transmitted signals start out, of course, as red, blue, and green information from the three picture tubes in the camera. The three signals are channeled into an encoder or transmitter matrix, and come out as luminance, I, and Q intelligence. These I and Q signals are each then put into a balanced modulator driven by the 3.58-MHz subcarrier oscillator (Fig. 12-6). The subcarrier is virtually suppressed during transmission and, along with burst, luminance, and sync, go to the transmitter for external broadcast. Greens are included in both I and Q, so no separate green signal is transmitted.

The modulation and demodulation processes are illustrated by a diagram from Sylvania's Color TV Clinic Manual, Volume 1, by Eugene Nanni and his staff at GTE Sylvania, Batavia, New York. We've made a couple of changes, however, to suit our immediate purposes. See Fig. 12-7. As I and Q information reaches the transmitter diodes, the 3.58-MHz subcarrier makes each conduct in the peak signal sequence, turning one diode off while the other is on—a time sharing process called synchronous detection so that both signals may be placed on a single carrier. While in the receiver, the I and Q signals are initially phase shifted 33 degrees by the receiver circuitry, and R-Y, B-Y (and G-Y in late receivers) is demodulated by an internally generated but burst sync'd 3.58-MHz oscillator to produce the same picture tones (phase and amplitude) of the transmitted signals. There are a number of ways to generate such synchronous demodulation which we cover shortly in the discussion of various receiver demodulator techniques.

Meanwhile, you should remember that chroma modulation is placing I and Q sidebands on two 90-degree out-of-phase suppressed carriers, removing luminance and transmitting video and color separately but on the same carrier. Conversely, demodulation is synchronous chroma sampling plus adding the luminance back to the chroma information by matrixing so there is brightness, including fine

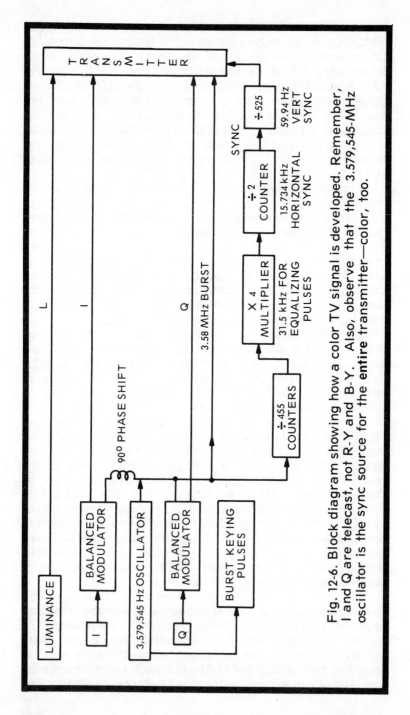

Fig. 12-6. Block diagram showing how a color TV signal is developed. Remember, I and Q are telecast, not R-Y and B-Y. Also, observe that the 3.579,545-MHz oscillator is the sync source for the entire transmitter—color, too.

detail, in all color scenes. Recall that in monochrome reception there are no I and Q signals processed and luminance **only** is distributed usually to the cathodes of the CRT, with just DC bias to the control grids. When chroma is transmitted, you need to recall that the I intelligence includes colors ranging from bluish-green (cyan) to orange, while the Q channel produces colors from yellowish-green to purple (magenta). Again, the bandwidth of the I signal is 1.5 MHz and the Q channel is 0.5 MHz. But the receiver's bandpass is such that it can demodulate color only to 0.5 MHz, whether R-Y or B-Y, and this is why the incoming chroma appears as a 1-MHz passband (3.08 to 4.08 MHz) across the 3.58-MHz reconstituted subcarrier (Fig. 12-1).

RCA'S CTC25 CHROMA CIRCUITS

In this mid- '60s vacuum tube receiver (Fig. 12-8) chroma is applied to the bandpass amplifier through terminal J, while terminal K carries the flyback turn-on pulse for the burst amplifier. With no incoming chroma, the potential at the V707B blanker grid is -90 volts, much of which finds its way to

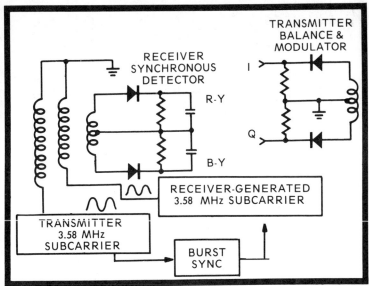

Fig. 12-7. Simplified modulation and demodulation diagram with subcarrier "sampling" diodes to produce colors at the conduction peaks. The transmitter 3.58-MHz oscillator syncs the receiver subcarrier oscillator by the sync burst.

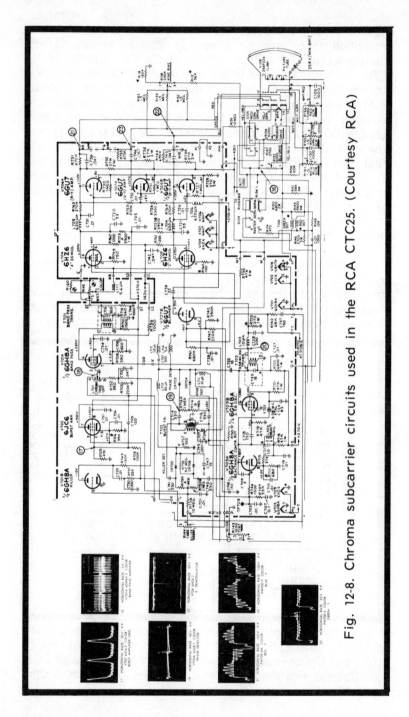

Fig. 12-8. Chroma subcarrier circuits used in the RCA CTC25. (Courtesy RCA)

terminal 1 of the R144 1-meg color killer potentiometer, while the top of this control, pin 3, sees +27 volts at the cathode and suppressor of the burst amplifier, showing that this tube is conducting because of the continuous flyback pulse. With no incoming color, the killer control is adjusted so that it just cuts the color killer on, delivering negative DC to the grid of the bandpass amplifier, cutting the tube off. With a chroma input, however, color sync reaches the burst amplifier through C705, and turning on the burst amplifier feeds burst transformer T702. The transformer output goes first to voltage-phase comparator diodes CR705 and CR706, which generate a varying control voltage to turn off the color killer so that its plate becomes zero, and permits the bandpass amplifier to receive chroma at its grid through chroma takeoff coil L701.

At the same secondary of the burst transformer are two more phase detector diodes, CR703 and CR704, that compare the phase of the 3.58-MHz oscillator with the phase of the incoming burst. The DC correction varies the grid voltage on reactance control V703A. This tube looks like a capacitor to crystal-controlled 3.58-MHz oscillator V703B, and so holds the oscillator in sync with the transmitted color sync burst by speeding up or slowing down the oscillator. Signals from the 3.58-MHz sine-wave oscillator go to pentode X and Z demodulator suppressors (the phase of the signal to V709 is shifted 90 degrees). In-phase chroma signals go to the V704-V709 grids. The two tubes conduct on the positive sine-wave tips, producing R-Y and B-Y outputs that differ in phase by less than 90 degrees. Green is resistively derived in the G-Y amplifier through the voltage divider between the X and Z demodulators and 270-ohm common cathode bias resistor R728. Chroma is then amplified and coupled by capacitor-shunted resistors to the CRT grids. Luminance is applied simultaneously to the tube's cathodes because of the 0.8 microsecond delay line between the second and third video amplifiers.

Arriving through PW700 terminal T is a large positive blanking pulse that drives the V707B blanker into inversion and produces a negative blanking pulse every 52.4 microseconds for some 11.1 microseconds duration, which turns off the three color amplifiers. In Fig. 12-8 the color amplifiers are passing a color-bar generated pattern which shows large negative off periods, with considerable variation in AC color-bar amplitudes, depending on color signals reaching the RGB (-Y) outputs. The DC plate voltages with no incoming signal can be almost identical.

In operation, the blue gun is conducting the hardest, with red second, and green only a few volts, since we might assume

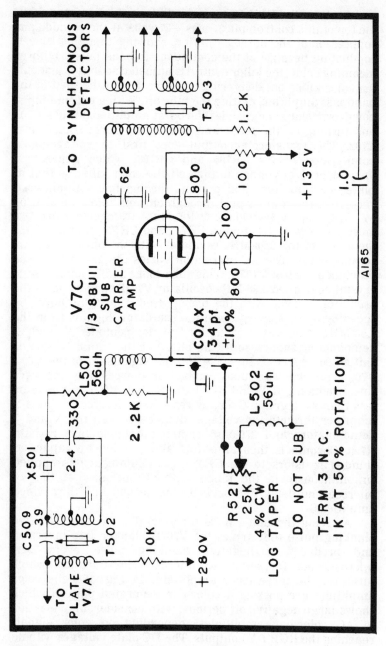

Fig. 12-9. General Electric's 3.58-MHz crystal ringing circuit. (Courtesy Bob Hannum)

the blanking pulse was included as part of the recorded waveforms, as it should be in the usual technician's evaluation. In engineering, we would mentally separate the two at once, since neither is a function of the other, and are completely separate entities. From the amplitudes of the various waveforms we could then deduce that blue was probably the weakest phosphor, followed by red, and then green.

GENERAL ELECTRIC'S CRYSTAL RINGING CHROMA OSCILLATOR

Another way to sync a chroma subcarrier oscillator is to use a high Q crystal filter (Fig. 12-9). This actually rings in damped sinusoids when "hit" by the color burst signal. Its advantage is that neither an automatic phase control (APC) nor a color killer circuit is required, since the crystal will not oscillate unless there is incoming color and, when excited, will directly provide all sync oscillations in the proper phase. Although the subcarrier amplifier shown in Fig. 12-9 is a tube type, the ringing crystal works equally well with semiconductors.

R521 is the tint control that induces a vector-like RL phase shift in the grid of V7C so that the best fleshtones can be manually tuned. Feedback is applied through the 1-mfd capacitor from the negative lower winding of T503, while the 800- and 62-pf capacitors in the grid and plate circuits are stray oscillation-preventing filters. Notice that the screen grid resistor is only 100 ohms, while the plate supply from the same DC source is applied through a 1.2K resistor and the primary of T503. This keeps the output at a certain level and "limits" the output to a certain DC swing. The voltage drop across either the 100-ohm or 1.2K resistors divided by the values of these resistances will, of course, tell you the current drawn by each electrode.

MOTOROLA'S DISCRETE & IC COLOR SUBSYSTEM

The Quasar CTV8 chassis (Fig. 12-10) contains an integrated circuit color processor (IC2) and an integrated color demodulator (IC1). IC2 is an MC1398 which provides DC control for both chroma amplitude and phase shift, a crystal-controlled feedback oscillator, built-in noise immunity, a Schmitt trigger color killer automatic chroma control, burst gate and pulse shaping, oscillator lock, and built-in voltage supply regulation.

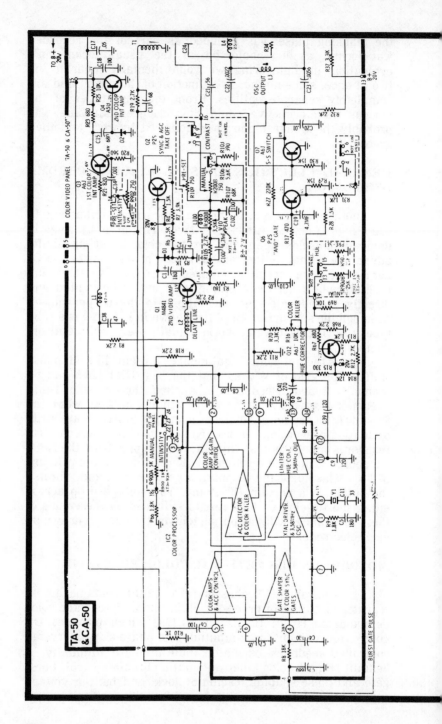

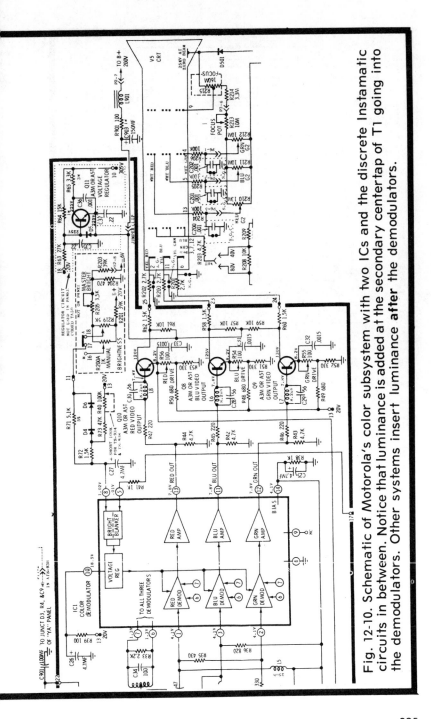

Fig. 12-10. Schematic of Motorola's color subsystem with two ICs and the discrete Instamatic circuits in between. Notice that luminance is added at the secondary centertap of T1 going into the demodulators. Other systems insert luminance after the demodulators.

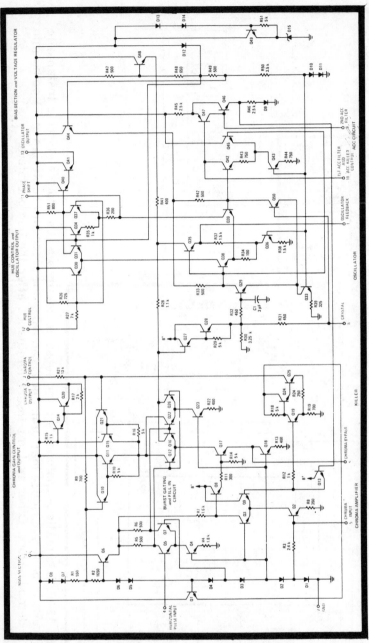

Fig. 12-11. Schematic of the Motorola MC1398 IC used in the CTV8 color processor. (Courtesy Bill Slavik)

A schematic of the MC1398 is shown in Fig. 12-11. Signal voltage for the color amplifiers and automatic chroma control functions comes from the emitter of the first video amplifier to pin 5, with a 0.05-mfd capacitor bypass at pin 6 (Fig. 12-10). The chroma amplifier (Q2, Q3, Q8, Q9, Q17 and Q18) takes its DC references from the series diode regulator string and upper left regulator output transistor Q6. The chroma amplifier supplies inputs for the chroma gain and output section directly above. Transistor Q2 is a constant-current source, while Q13 is a bias and operating source for both the chroma amplifier and the color killer Schmitt trigger to the right of the amplifier. Differential amplifier Q3-Q9 supplies Q8 with incoming chroma that is emitter-coupled to the base of Q17, another amplifier which receives regulated emitter current by bypassed Q18. Output from inverter Q17 now goes to the emitters of Q12 and Q16, half of a dual differential, constant-current fed (by Q23) Q22 and Q26 gated switch.

Gating for the differential amplifier (applied through Q5 and Q7) is switched by a 4-microsecond 3-volt p-p flyback pulse to pass the reference burst information used to lock the 3.58-MHz chroma oscillator. The second half of switched amplifier Q22-Q26 is used to prevent undesirable level changes in chroma during gating.

Outputs now go to chroma gain control differential output amplifier Q11-Q15 and the gating inputs Q10 and Q21. DC bias for both Q10 and Q21 is set by the chroma **intensity** control outside the IC by voltage division from the internal connection between R1 and R2. Therefore, the conduction of Q11-Q15 is controlled and Darlington output Q14 and Q20 provides a linear chroma signal at pin 2. Without incoming chroma, there is no bias from Q13 for the Schmitt trigger, so Q19 goes high while Q25 conducts hard, pulling the base of Q21 down and turning the transistor off so there is no noise or transient output.

The oscillator, of course, generates the 3.579,545-MHz chroma subcarrier, with a conventional crystal attached to pin 8. An RC series feedback network connects to ground from pin 1, a hue corrector and color killer control circuit at pin 10, and another capacitive filter at pin 9. Darlington pair Q27 and Q28 pick up the burst signals from Q16 and Q26 in the chroma amplifier section and deliver them to the base of the first section of the subcarrier oscillator, Q29. This transistor and Q50 form a differential pair from which the output level is proportional to that of the incoming burst. The conduction of each transistor then excites the rest of the Q35 through Q39 oscillator network. Oscillator output through Q38 goes to the emitters of Q30-Q31 so that hue control can be established. An oscillator signal also is applied to the emitters of a second differential pair, Q34 and Q37, for a 3.58-MHz output through Darlington pair Q40 and Q41 to pin 13—the oscillator signal exit for all three demodulators. The sine wave from Q39 is phase shifted somewhat by R36 and external capacitor C9. The hue control is simply another DC bias potentiometer that can manually change the bias on the two differential amplifiers which are already 180 degrees out of phase. DC bias changes mix the contents more or less and change the resulting output phase.

The automatic chroma control circuit supplies two operations: the first rectifies the subcarrier sine wave by differential offset amplifier Q42-Q45 and puts out a DC voltage proportional to the incoming chroma signal. The second function is adjustment of the operating point for Q46 and Q47 through the action of the controls at pin 10. With incoming chroma, the ACC output turns on the Schmitt trigger input, Q19, and turns off output Q25, which then has no affect on Q21 and Q15, but Q13 still supplies operating bias for Q9. With no ACC, naturally, Q19 is off, clamp Q25 is on, Q9 has no base bias and so ceases to conduct. Voltage regulation and bias drops are generated by Q48 and Q49 with D12 through D14 and reference zener D15, while D10 and D11 provide base bias for Q33, the current feed for Q29 and Q50.

Normal chroma out (Fig. 12-10) then passes through pin 2 to mixing transformer T1, while luminance information arrives at the centertap in the T1 secondary, where both chroma and luminance are combined. But, remember, the chroma must be demodulated. The 3.58-MHz subcarrier sine-wave output appears at IC2 terminal 13 and is applied by way of RLC coupling to the red, blue, and green demodulators. In passage to the demodulators, however, the chroma information meets some unfamiliar passive and active components making up the "Instamatic" circuits.

There are two groups of circuits in the Instamatic: one, the variable potentiometer group which is preset and which controls the brightness, hue, contrast and intensity. An active semiconductor group increases the red output gain and monitors the amplitude of the second color "Int." amplifier and changes the demodulator phase. Transistors Q6 and Q7 act as an AND gate when the Instamatic is active, so that Q7 shunts the red drive control (R32) and allows more current to flow in the Q10 emitter, and therefore, the collector. As a result, Q10 conducts harder and supplies a slightly reddish tint to the CRT screen. At the same time, the 3.58-MHz phase angle to the blue demodulator is changed slightly through C21 and, together, the two changes cause the screen to broaden the area of fleshtones.

The other active part of the Instamatic circuit is composed of color "Int." amplifier Q3 and Q4. A chroma signal applied to the base of Q3 where it is amplified, then the chroma goes through C15 and negative rectifying diode D2. This forms a relatively linear AC-to-DC feedback which controls the intensity of the color amplifier. The feedback is applied to pin 3 on the IC processor when the Instamatic switch is turned on.

Luminance information passes through delay line L2 to the base of the second video amplifier. Two video signals are taken from Q1, one from the collector for the noise limiter and emitter-follower sync and AGC gate, and the other goes from the emitter to a series peaking coil and RC video peaking to the secondary centertap of T1. A negative spike of voltage (noise) strong enough to turn on D1 causes the diode to shunt both the AGC and sync outputs to AC ground through C2. Limit adjust potentiometer R74 (not shown) sets the anode bias for D1.

MOTOROLA'S TRI-PHASE CHROMA DEMODULATOR

The chroma IC demodulator (no number as yet) is a thoroughly unique device in that it mixes luminance with the R-Y, B-Y, G-Y before demodulation. Pure RGB composite chroma and luminance are supplied to the final RGB amplifiers.

The circuit in Fig. 12-12 accepts combined luminance and undemodulated color signals from T1 (pins 6 and 7) 180 degrees out of phase—for each of the red, blue, and green processors. The subcarrier sine wave is supplied directly from IC1 pin 13 by LC coupling to the three demodulators at the required phase relationship to produce properly phased outputs. The brightness control (R200A) sets the operating potential on the brightness-blanker stages that, together, control a pair of Darlington (Q9-Q10) current multipliers to supply collector voltage and current for the trio of virtually identical demodulators. The DC input is pin 10.

Positive chroma goes from the top winding of the T1 secondary through demodulator pin 7 to the base of Q8 (the driver-current supply for blue demodulators Q3 and Q4), to the base of Q18 (the green drivers Q14 and Q15) and finally to Q25 (the keyer for red demodulators Q21 and Q22).

Negative chroma comes from the bottom of the T1 secondary, 180 degrees out of phase, and is supplied to the bases of current drivers Q7, Q17, and Q24 for the blue (Q1 and

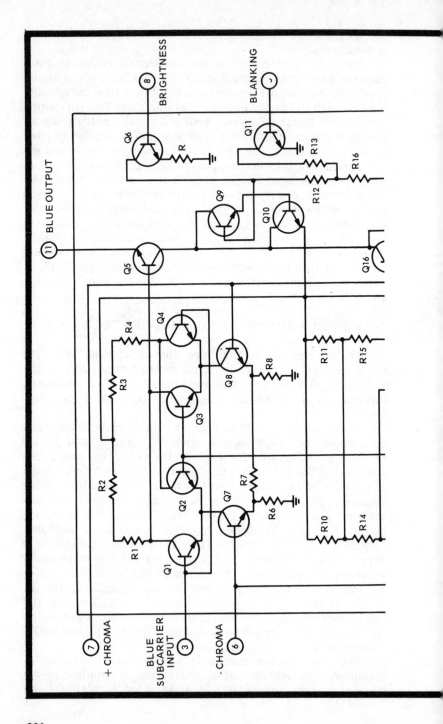

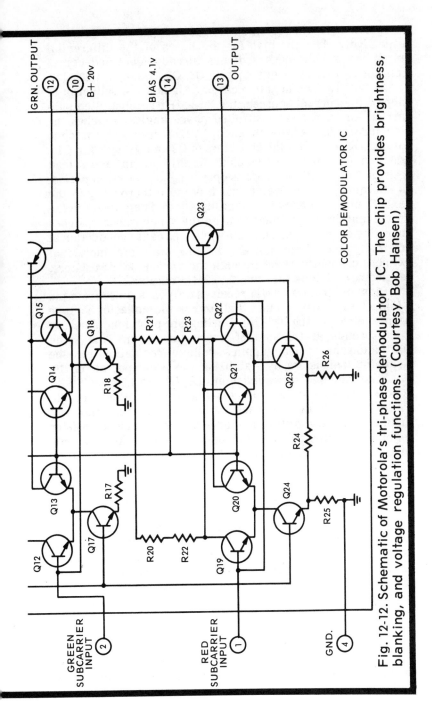

Fig. 12-12. Schematic of Motorola's tri-phase demodulator IC. The chip provides brightness, blanking, and voltage regulation functions. (Courtesy Bob Hansen)

301

Q2), green (Q12 and Q13) and red (Q19 and Q20) demodulators, respectively. The DC bias for the bases of the differential amplifiers' synchronous switches already identified comes from the external 20-volt supply to Q2, Q3, Q13, Q14, Q20, Q21.

As positive and negative chroma signals are delivered to each of the current sources, the 3.58-MHz reconstituted subcarrier sine wave (at controlled phase angles) reaches the demodulator bases through pins 3, 2, and 1, reading from top to bottom. Observe that the collectors of Q2 and Q4, Q12 and Q14, Q20 and Q22 are crosscoupled directly to an extra load resistor, since only one gate is operating at any given period.

Chroma signals (we discuss this more thoroughly in the next IC analysis) are applied positively and negatively to the six current driver supplies to keep the synchronous switches operating as the subcarrier sine wave tops cause conduction at externally set phase angles. In this way the individual switches conduct current through the various load resistors, developing voltages that are 7.16 MHz outputs.

Blanking transistor Q11 receives a flyback pulse from pin 5 that clamps it to ground and lowers the collector voltage supply through the Q9-Q10 Darlington pair for about 10 microseconds at the end of every line scan. This same Darlington pair also supplies voltage regulation for the switching transistors by stabilizing the B+ input at pin 10. Capacitor shunted series inductors L6, L7 and L8 (Fig. 12-10) oppose any subcarrier voltages that might filter through, while the RGB drivers each have a DC current flow adjustment in the emitter. A diode-clamped single transistor regulator supplies collector voltage for the three RGB drivers. Individual color signal outputs go directly to the CRT cathodes.

ZENITH'S THREE-CHIP CHROMA PROCESSOR

For an idea of what these three chroma processors look like when mounted in their plug-in sockets, see Fig. 12-13. Fig. 12-14 is a schematic of the thick film resistor matrix interface between chroma and subcarrier regenerators. There are no transformers to twiddle in these chroma circuits, and only three DC pots that are quickly adjusted for complete chroma alignment. However, instead of simply eyeballing the adjustments with a color-bar pattern, we would recommend that the crosstalk adjustment be done with a vectorscope, too. You will see why later. Careful adjustment will allow maximum bandpass and chroma input amplitude—two essentials for quality color reproduction.

The uA780, 781, and uA746 (Fairchild designation) constitute an automatic phase control loop chroma processor in three monolithic integrated circuits that have a low phase error because of the high-sensitivity phase locked loop. The ICs are housed in 14 and 16-pin hermetically sealed ceramic packages for easy insertion and removal from commercial IC holders. In volume, all three collectively should be available this year (1972) for $2.

uA781C Chip

The 14-pin chroma processing chip is a uA781C Fairchild monolithic IF amplifier consisting of a pair of complex gain-controlled circuits (Fig. 12-15) in series, with the first designated as an automatic chroma control amplifier which receives all chroma inputs, but at all times is subject to DC ACC clamp from the uA780C IC chroma subcarrier regenerator. This amplifier is capacitively coupled to the chroma level amplifier, which, in turn, is clamped at its own DC swing by the chroma level control. Amplified chroma output goes to the uA746 third IC chip for color demodulation. Between the ACC amplifier and the chroma level amplifier is a color killer switch with another DC killer adjust that tells this unit at what level to kill the chroma output. The actual

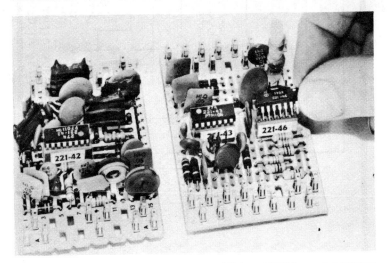

Fig. 12-13. Monolithic ICs in Zenith's 19CC19 and 25CC55 hybrid and solid-state color chassis. These three chips do all the chroma processing, and are hand-removable from their mounting sockets. (Courtesy Ed Polcen)

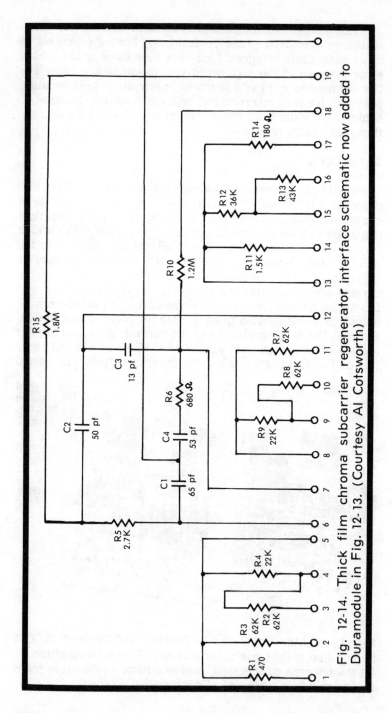

Fig. 12-14. Thick film chroma subcarrier regenerator interface schematic now added to Duramodule in Fig. 12-13. (Courtesy Al Cotsworth)

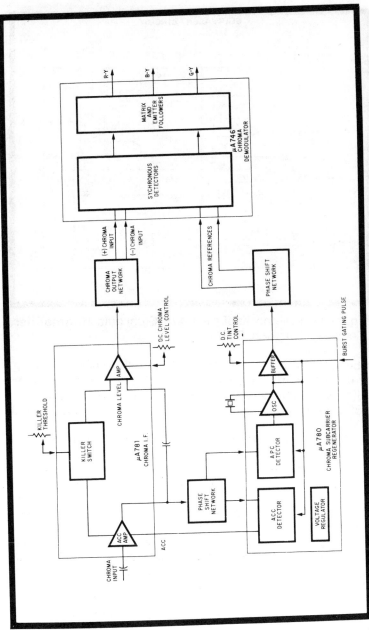

Fig. 12-15. IC manufacturer Fairchild's block diagram of the Zenith-developed three-chip processor. (Courtesy Don Smith)

305

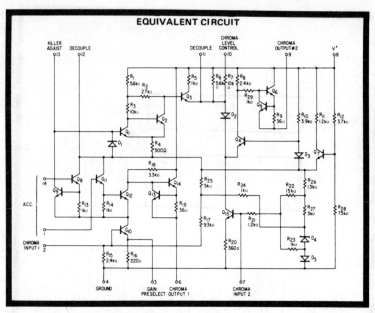

Fig. 12-16. Schematic of the uA781C chroma IF amplifier. (Courtesy Norman Doyle)

schematic for the uA781C chroma IF amplifier is shown in Fig. 12-16. ACC inputs are pins 1 and 14, which connect to emitter-follower Q11 and decouple stage Q8. Chroma comes in at pin 2 and goes to the base of Q10 for initial gain-control amplification.

DC voltages from the subcarrier regenerator are simultaneously applied to the automatic chroma control inputs. The potentials are different, however, and when Q8, for instance, is driven further into conduction to supply more base current to Q9 (which makes this emitter follower furnish more collector-emitter voltage for Q10 and Q12), the opposite phase voltage on Q11 causes this emitter follower to drive Q12 less. As a result, the output of Q12 is regulated, as well as the output of follower Q14, the first chroma output shunted by switch Q13. An opposite condition can occur, too, where Q11 can drive Q12 harder when there is less collector-emitter voltage from Q9. Therefore, the gain-controlled chroma output at pin 6 is coupled to the input of the second chroma amplifier (Q15) at pin 7.

DC for the base of Q15 is regulated by zener diodes D4 and D5, which are supplied by Q7 from the power supply. D4 and D5 also furnish collector voltage and emitter-base bias for

Q10, Q1, Q2, Q11, and Q8 in the stages just described and in the Q1, Q2, Q3 DC gain control circuits that monitor the DC from the level control relative to the potential at the collector of Q9 and the killer adjust. Diode D2 is a gain control, operating together with Q4 (like Q9 and Q12). Chroma from the collector of Q15 forward biases the emitter of Q4 and reaches the base of emitter follower Q6 as a positive-going signal identified at pin 9 as chroma output 2. Diode D3 supplies both bias and temperature compensation for amplifier Q4.

The block diagram says that the killer adjust affects both sets of chroma amplifiers, and indeed it does, for when there is no incoming color, and, therefore, no ACC input, Q8 and Q9 cease conduction, there is no chroma no. 1 output and, consequently, no chroma no. 2 input or output. Actually, Q1 and Q2 form a Schmitt trigger, and as long as Q1 is off, Q2 is on, and Q3 can't operate, since its base is pulled below turn on. While automatic chroma is incoming, Q1 and Q2 remain in this condition and the circuit continues in full operation. But when the "kill point" is reached, Q2 turns off, forward bias appears across R2, Q3 turns on, and all current in Q15 is shunted through D2 to AC ground. Consequently, the output circuit is completely cut off. Short circuit protection is provided by Q13

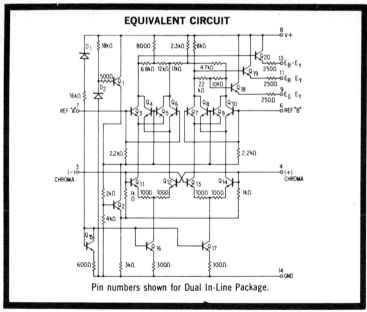

Fig. 12-17. Fairchild's uA746E synchronous demodulator IC schematic used by Zenith. (Courtesy Frank Hadrick)

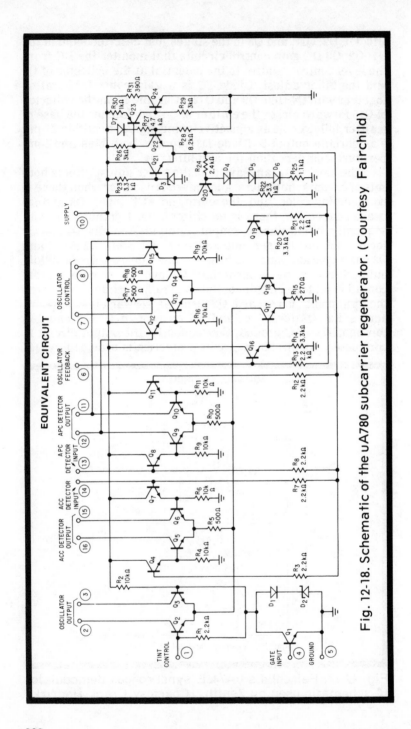

Fig. 12-18. Schematic of the uA780 subcarrier regenerator. (Courtesy Fairchild)

at pin 6 and Q5 across Q6 at pin 9, these being the chroma 1 and chroma 2 outputs, respectively.

uA746E Chip

The IC demodulator chip (Fig. 12-17) has 29 resistors, 20 transistors, and 2 zener diodes that, along with three diodes, supply ample regulation for the two pairs of differential amplifiers and the eight emitter cross-coupled demodulator switches. Depending on whether the cathode ray tube red phosphor is gadolinium or yttrium, these ICs, through the phase-shifted input networks, will demodulate at approximately 90 or 105 degrees, respectively. We say "approximately" because modern CRT phosphors are somewhat different from those originally specified by the NTSC, and even though efficiences are now approaching relative equality, slightly varying phase and drive adjustments are still needed to produce the best fleshtones (and Kelvin temperatures) without seriously neglecting a desirably true chroma reproduction of the surrounding scenery.

The uA746E is a network of transistors and resistors formed on a single substrate (foundation) so that the entire chip is monolithic—that is, a single piece of silicon. Continuous wave (CW) 3.58-MHz signals from the regenerated subcarrier oscillator are fed into REF "A" and REF "B," while positive and negative chroma goes to the pair of differential amplifiers that key the eight switching transistors in the differential collectors. Since a CW sine wave is always present, it is the differential amplifiers that control the sequence of operation. Conduction of the differential amplifiers is proportional to the incoming chroma signals. This, in turn, determines the output of the synchronous demodulator switches above them.

If negative chroma information is arriving at terminal 3, it is coupled to the bases of Q11 and Q13, and if plus chroma information is received at terminal 4, it travels to the bases of Q14 and Q12. Depending on the amplitude, Q11 or Q12, and Q13 or Q14 conduct more or less, turning on partially or fully the associated switches. If Q11 and Q12 are in conduction, the emitters of Q3 and Q5 are enabled by Q11 and the emitters of Q4 and Q6 are turned on by Q12. The 3.58-MHz subcarrier sine wave, meanwhile, forward biases the bases of Q3 and Q6. Since 3.58 MHz signals put to the bases of opposite, collector-coupled transistors 180 degrees out-of-phase, and at 7.16 MHz, the chroma from the differential amplifiers is demodulated, but the subcarrier signals are virtually cancelled at the collectors of Q3 through Q6, thereby removing the need for extensive external filtering, except for an LC circuit following the output load resistors. The same sequence of operations is true for amplifiers Q13 and Q14, and switch demodulators Q7 through Q10.

The intelligence from these two pairs of double-balanced demodulators is matrixed resistively in the respective collector circuits, with the outputs going to Q20, the B-Y emitter follower, Q19, the R-Y emitter follower, and Q18, the G-Y emitter follower. Observe that the green follower tapoff is between the 22K and 10K collector resistors in the Q7 to Q10 switch group. The G-Y is derived by adding parts of the inverted B-Y and R-Y chroma inputs into Q13 and Q14.

Biasing for current sources Q16 and Q17, which feed the two pairs (Q11 through Q14) of differential amplifiers, is supplied through zener D1 and the 16K series resistor. Diode-connected Q15 is fed by the same source. The Q15 emitter is above ground by 600 ohms. The bases of Q16 and Q17 are clamped by this regulated voltage and supply a constant current for the amplifiers. A 3-volt bias for the bases of differential amplifiers Q11 through Q14 originates from Q1, whose base is clamped at a fixed potential by zener D2. Initially, the emitter voltage of Q1 is 6 volts, and this is applied to the bases of the eight switching transistors. A voltage divider, however, drops this value at the base of Q2 to the point where the emitter of Q2 supplies a 3-volt bias, which is shunted to the bases of Q11 and Q14 through 1K resistors to fix these potentials at a positive forward bias of slightly less than 3 volts.

uA780 Chip

The uA780 phase locked loop subcarrier regenerator contains the 3.58-MHz oscillator that is phase-locked to the transmitter by burst through the automatic phase control detector. This IC also furnishes burst gating, DC tint, and automatic chroma control, plus its own regulated voltage supply. The schematic is shown in Fig. 12-18. An overall interconnection diagram for the entire chroma processor appears in Fig. 12-19.

The oscillator itself (pins 6, 7, and 8) is made up of two pairs of differential amplifiers, with bias and control gates at all inputs. Rigid control is provided by the APC detector DC outputs at pins 11 and 12. The oscillator, through feedback out of the base and emitter of Q16, drives the Q17 section of differential amplifier Q17-Q18 to supply the emitters of the APC and ACC detectors as well as the oscillator output at pins 2 and 3. The oscillator output signals are phase shifted by RLC components in the ladder network phase shifter and delay line where the sectional or total time delay can be calculated by N (the number of sections) times the square root of L and C: written:

$$Td = N\sqrt{LC}$$

while the output impedance is:

$$Zo = \sqrt{L/C}$$

This network is strikingly reminiscent of the DC delay lines used in radar for square wave pulses, but used here basically

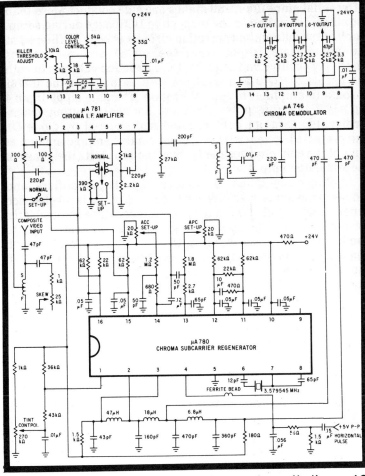

Fig. 12-19. Interconnection diagram for all three IC monolithic chips in the complete Zenith chroma processor. (Courtesy Bill Edlund, Fairchild)

to separate the phase of the voltages going to pins 6 and 7 of the uA746 chroma demodulator by approximately 105 degrees.

APC and ACC setup controls are simple 20K potentiometers which are adjusted with the normal-setup switch in the setup position to provide bias at the bases of Q8 and Q7, respectively, without the benefit of incoming chroma. The subcarrier oscillator crystal is connected to pins 6, 7, and 8 (oscillator control), with feedback going to Q16 through pin 6 to complete the regenerative oscillator loop. Output from amplifier Q17 determines the phase for both the ACC and APC differential amplifiers, and makes one side conduct more than the other, depending on the respective DC biases during setup. Each of the automatic correction circuits is heavily filtered at the appropriate output from the uA780 and so supplies relatively positive and negative DC correction voltages to both the oscillator (Q12 and Q15), as well as pin 1 of the uA781 chroma IF amplifier to control its gain. The gate input at pin 4 (Q1) is a blanking pulse from the horizontal output transformer, temperature compensated and shunted by diode D1 and protected by zener D2. A DC tint control voltage applied at pin 1 of the chip changes the Q2 bias, and so shifts the phase of the oscillator output.

Voltage regulation is achieved by the amplifiers and diodes in the upper right corner of Fig. 12-18. D3 is a reference zener for Q21, while D7 is a feed and temperature compensator for the base of Q22. Supply voltage changes will cause Q22 to conduct more or less, changing the base voltage of Q23 which conducts proportionately. As a result, Q24 varies the load on the 24-volt power supply, providing dynamic regulation for this portion of the chroma subsystem. D4 through D6 are also bias and temperature compensators for emitter follower Q20.

The three chips are manufactured at least by RCA and Fairchild, and perhaps others. RCA numbers for the Zenith system are: CA3070, CA3071, and CA3072.

RCA'S TWO-CHIP CHROMA PROCESSOR

The two-chip chroma processor for this receiver is shown in Fig. 12-20 with output connections to the MAD CRT driver modules that are the encapsulated thick film throwaways exhibited in Fig. 12-21. The MAC chroma 1 module is shown as a pair of "amplifiers" included in the A and B subsections of IC1, with outputs going to MAE chroma module 2, which has a voltage regulator governed by the output of pin 6, IC1B of chroma 1. All areas within the smaller dashed lines, of course, are on edge-connector plug boards, while the large dashed

312

lines denote the mother board for each particular section. This series of RCA solid-state receivers has four mother boards, 12 plug-in modules, and a removable, single-piece high-voltage quadrupler.

We'll look first at the IC1 CA3066 (Figs. 12-22 and 12-23), that contains most of the color processors except the tint control and demodulators. Signals from the video preamplifier are coupled to pin 1 of the chroma amplifier. This stage, in which gain is automatically controlled by the ACC detector amplifier, supplies both the burst and bandpass amplifiers with signal. The chroma amplifier is stagger tuned to 4.3 MHz at pin 16 by the variable inductor and RC network, while the bandpass amplifier is tuned to 2.7 MHz, pin 13, by L2 and the RC combination so that the two combined can pass the 3.08- to 4.08-MHz chroma signals.

The subcarrier signal is introduced between pins 7 and 11 by a crystal ringing circuit. This type of oscillator control is a somewhat less expensive method of color sync than the phase comparator, and the neutralization and damping of the circuit must be carefully designed for best transient response, especially so that no hue shading appears after the vertical interval. On the other hand, with this type of circuit you could dispense with the color killer, but RCA didn't. You can drive the ACC with the 3.58-MHz oscillator, since the amplitude of the ringing signal is proportional to the amplitude of the burst, and a well-adjusted ringing circuit will lock on a single cycle of burst. The circuit also must be followed by a well-balanced demodulator, and it **does** need a color killer **if** automatic color control is a feature, which it is.

A horizontal keying pulse gates chroma from the bandpass amplifier into the burst amplifier, and this stage, of course, rings the crystal for the 3.579,545-MHz subcarrier oscillator. Chroma gain is simply another potentiometer, as are the ACC adjust and color killer bias setting. Inductors L1 and L2 need alignment with a sweep generator for both bandpass and gain.

Fig. 12-23 shows the schematic diagram of the IC1 color processor. Common-emitter and current source amplifier Q1 accepts chroma for collector coupling into the emitters of differential amplifier Q16 and Q2. The Q16 base is controlled by the amplitude of the 3.58-MHz oscillator through ACC action, and the output of Q2 is tuned by L1 and coupled through Z2 into the base of emitter-follower buffer Q19. This stage directly supplies Q3 and resistively controls the base of Q6 that, together, furnish current for differential amplifiers Q21-Q4 and Q7-Q24, the initial bandpass and burst amplifiers.

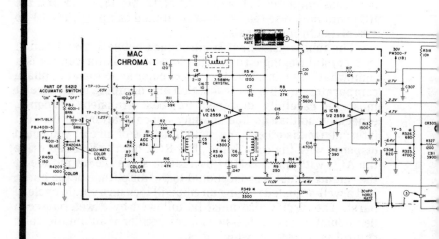

Fig. 12-20. RCA's sophisticated all solid-state CTC46 two- chip chroma processor schematic. (Courtesy Tom Bradshaw, George Corne)

Burst keying appears at terminal 10 and is applied from emitter follower Q25 to the bases of Q7 and Q21 at a pulse width of 5 microseconds and 8 volts in amplitude. DC chroma gain adjust for Q20 and the bases of Q5-Q22, Q4 and Q24 is applied at terminal 15. Q5-Q22 is the final differential bandpass amplifier with its output at terminal 13, while Q7 is the burst amplifier output to L3, the crystal ringing transformer. The chroma bandpass amplifier is tuned by L2 and passed through zener Z6 to the chroma output at terminal 14, then to the demodulators.

Burst rings the crystal whose phase and frequency control the DC-coupled group of 3.58 MHz oscillator transistors connected to Q11 following terminal 7. The output of the sub-carrier oscillator goes to terminal 8 and also through the

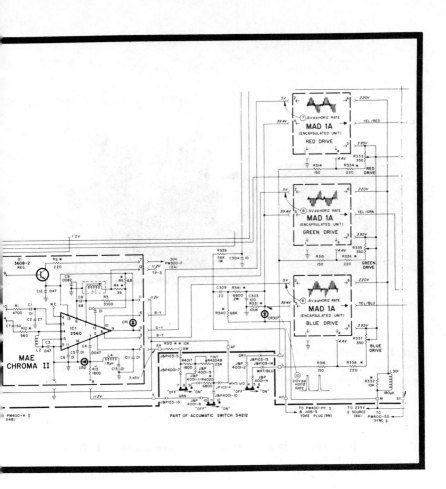

emitter of Q8 to the base of Q26. The oscillator is DC biased by ACC adjust R2. This is a loop gain adjustment where Q8 supplies current for the killer amplifier, in which the level is set by 47K R1, shunted by a 100-mfd capacitor. Q27 supplies DC for Q12 to keep it operating with the higher levels of the sub-carrier oscillator, so that Q13's base is receiving little or no voltage and is probably not conducting.

With no incoming burst, Q12 is off, and Q13's base rises to the bias voltage output of Q18. Q13 conducts and clamps the chroma gain input through pin 15 close to ground so that there is no bandpass amplifier output at all. Diodes D5 and D6 are probably temperature compensators, while Z3 is a voltage clamp. Zeners Z4 and Z5 connect to the 11- and 30-volt lines in these modular receivers and should be used as a constant

Fig. 12-21. Interior of MAD module from its beginning as a discrete, to thick film on ceramic substrate, then final encapsulation. (Courtesy RCA)

source for IC1, since their regulation is adequate and both track well with temperature.

The CA3067 chroma demodulator and tint control is another 16-pin IC, this time centered on the MAE chroma II plug-in module (Figs. 12-24 and 12-25). Subcarrier reference input is fed from terminal 3 to the tint amplifier. The tint amplifier gain is adjusted by the 25K DC control connected to terminal 2. Q7 base is at AC ground through a 0.01-mfd capacitor at terminal 16. Q2-Q3 is a differential amplifier, with Q3 base-fed from the terminal 3 input. The output of these two amplifiers is mixed in the collector of Q4, and the gain of Q4 and Q3 is adjusted by the tint control for an oscillator mix that results in a certain phase shift for desirable fleshtones. Q8, meanwhile, supplies DC bias for Q4 from the divider network of R2, R5, etc., that extends to the bases of Q6, Q5, and Q1 from the main 11.2-volt supply. Voltage from the Q4 collector passes to the base, then emitter, of Q38 and into the base of differential amplifier Q39-Q7 and out to the RLC phase shift

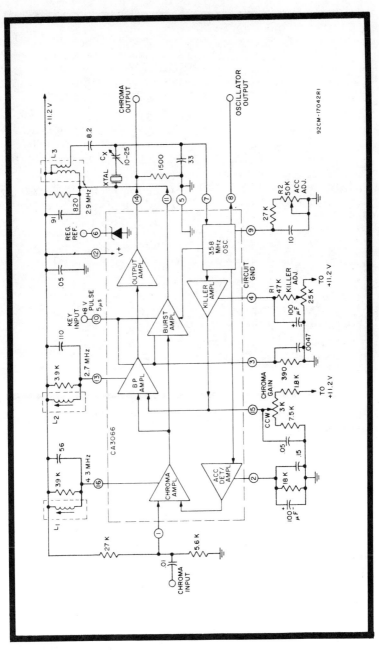

Fig. 12-22. Block diagram of the CA3066 chroma signal processor. (Courtesy RCA)

317

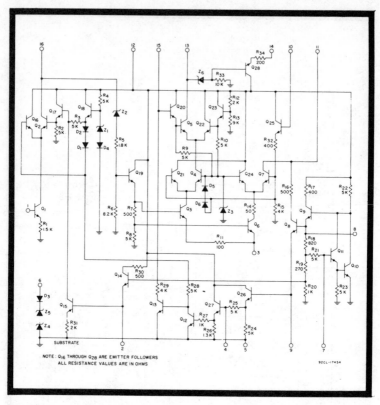

Fig. 12-23. Schematic diagram of the CA3066 chroma processor, IC1.

network for the amplifier-demodulator input terminals 6 and 12. These passive components determine the phase angles of the sine waves going into the amplifier-demodulators. Terminals 7 and 11 connect to additional emitter bypass filters.

Terminals 6 and 12 may be a little difficult to find, but they are located in the base inputs of Q9 (lowest amplifier) and Q11 (top amplifier), respectively. These two amplifiers are identical and seem to be self-adjusting, since the greater the input the more voltage dropped across R13, for instance. The greater the drop across R13, the less the base input to Q45, and the less signal appears through the emitter of Q44. The same is true for the Q9-Q40 group at the bottom of the schematic. The Q10-Q42 assembly is virtually a regulator for the base bias voltages of Q12, Q19, Q26, Q27, Q34, Q31, and Q28 so that when

318

the main supply changes, Q43 feeds back to Q10, and the DC outputs at the emitter of Q42 is automatically limited or boosted, depending on the drop across R19.

Chroma is applied to the lower Q13-Q14 and Q20-Q21 pairs of differential amplifiers, but only through terminal 14, since terminal 15 is at AC ground and the resonant frequency of the 620-microhenry coil and 0.01-mfd capacitor in series amounts to 63.7 kHz, with the reactance of the capacitor amounting to 1 ohm at 3 MHz. To describe the switching action of the synchronous demodulators: when Q13 and Q20 are receiving chroma, R18 and R28 develop sampling signals that key Q22-Q24 and Q15-Q17. And when R14 and R24 develop subcarrier references and turn on Q23-Q25 and Q16-Q18, synchronous switching develops that produces the R-Y and B-Y outputs from the top and bottom output demodulators which go to Q51 and Q47. The G-Y information is resistively matrixed by R37 and R40 for the Q31-Q33 output amplifier with a Darlington Q48-Q49 driver at its input. Capacitors C2, C3 and C4 are the 7.2-MHz filters that absorb switching transients. Since there is signal inversion in Q9 and Q11, signals to the bases of each pair of synchronous switches are 180 degrees out of phase with those opposite, so one pair switches while the other pair is off.

It's interesting to note that in the Q29-Q30, Q32-Q33 and Q35-Q36 output amplifiers, a bias is developed for Q29, Q32, Q35 in the collectors of Q30, Q33 and Q36. The bias keeps the lower transistors turned on and helps deliver a pair of reinforced, low-impedance, in-phase inputs to the MAD kine driver modules. These MAD modules are a trio of two-transistor amplifiers that mix chroma and luminance for RGB cathode drives of the picture tube.

The Accu-Matic in this receiver does about what most other fixed bias systems of this type do: it reduces the blue output, changes the phase angle slightly towards open, adding some red, and enlarges the demodulation angle so that the area of fleshtones is increased.

A ONE-CHIP COMPLETE MONOLITHIC CHROMA PROCESSOR

At the June, 1971, IEEE Broadcast & Television Receiver group meeting in Chicago, Fairchild showed a complete one-chip color processor. The finished product has a power dissipation of 400 mw, an output resistance of 50 ohms, five alignment adjustments, and 42 external components as opposed to the announced competition which has 62 and 77, respectively, for active 2- and 3-chip circuits. The system is identified as a uA782 chroma processor. A block diagram is

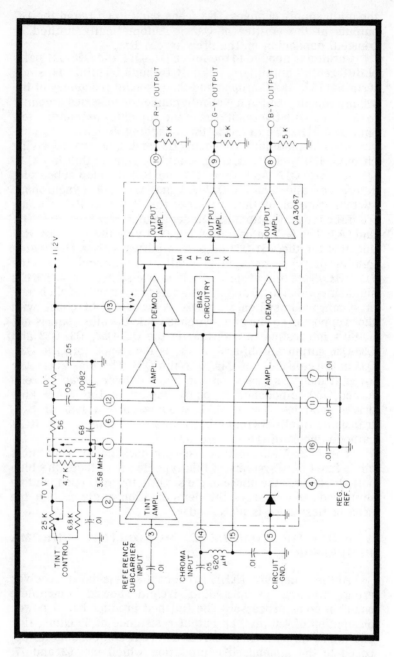

Fig. 12-24. RCA's CTC46, 54 Series CA3067 chroma demodulator functional drawing.

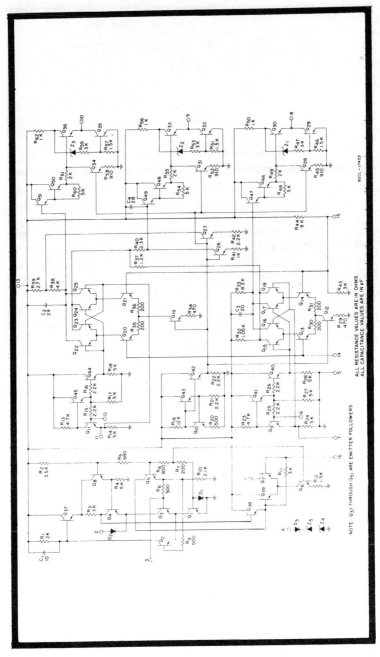

Fig. 12-25. Schematic of the CA3067 chroma demodulator.

321

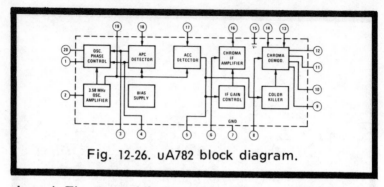

Fig. 12-26. uA782 block diagram.

shown in Fig. 12-26, while Fig. 12-31 outlines the entire function in what is described by Norman Doyle as a typical application. The external transistor is a blanker for 3.58-MHz reference signals to the demodulators. Other peripheral components include APC and ACC potentiometers, capacitance couplers and filters, the 3.578545-MHz crystal control for the color oscillator, video input and chroma reference input transformers. The block diagram (Fig. 12-26) starts with the 3.58-MHz oscillator that supplies the demodulators through pin 19, and the oscillator phase control, ACC and APC detectors, along with an active bias supply. The ACC detector output goes to the IF gain control and color killer, and these two stages act on the chroma IF amplifier and the chroma demodulators, respectively.

Color IF Amplifier (Fig. 12-27)

The basic chroma amplifier comprises two cross-coupled pairs, Qa-Qb, and Qd-Qe, with incoming video applied to the

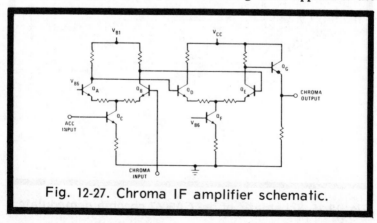

Fig. 12-27. Chroma IF amplifier schematic.

base of Qb from the video detector and the bandpass transformer at an amplitude of from 10 to 200 millivolts p-p. Automatic chroma control voltage at the base of Qc controls the DC gain of both transistors, with outputs to the bases of Qd and Qe, another differential amplifier supplied by constant current source Qf. Emitter follower Qg is the output with AC at about 2 volts p-p, and this voltage goes into the demodulators through an output that contains both level and tint controls.

Subcarrier Regenerator and Control Circuits (Fig. 12-28)

Again, a current supplying transistor Qc activates the emitters of Qa and Qb, which supplies the 3.58-MHz crystal, its tuning capacitor, and the base of Qd. Qd and Qa form a voltage-controlled oscillator. The oscillator frequency is determined by the voltage difference between the bases of Qa and Qb, which results in drops across R1 and R2 in the collectors of Qa and Qb, and the charging 56-pf capacitor between them.

Upon horizontal scan, the collector of Qd rises toward Vcc, back-biasing the emitters of Qe, Qf, Qg and Qh, and allows oscillator current to flow into the external LC circuit, supplying a pair of reference signals for the demodulators. But when burst appears, the slightly delayed negative horizontal cutoff pulse backbiases the "diode," and the oscillator current channels into automatic phase control detectors Qe and Qf and

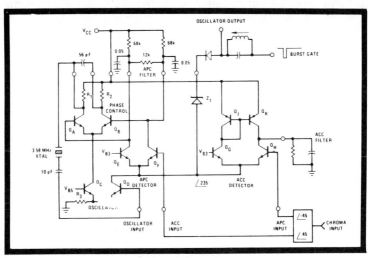

Fig. 12-28. 3.58-MHz subcarrier oscillator, phase control, APC detector, and ACC detector circuits.

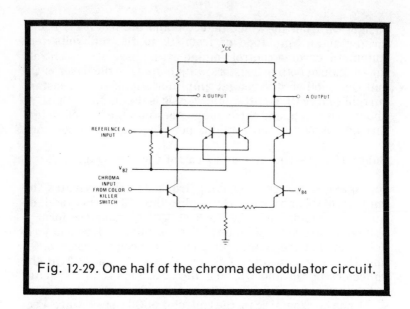

Fig. 12-29. One half of the chroma demodulator circuit.

automatic chroma control detectors Qg and Qh. Since plus or minus 45-degree difference phases of burst from the chroma input appear at the bases of Qf and Qh, Qf is turned on and its output goes to Qb and the APC filter to supply any correction voltage needed by the color oscillator. Since the ACC detector is turned on simultaneously with the APC detector, base-coupled Qk and Qj form a balanced-to-unbalanced converter, with initial collector currents equal. The output goes into an ACC filter for proportional DC control of the IF amplifier and color killer circuits.

Chroma Demodulator (Fig. 12-29)

This type of circuit is simply a lower chroma input stage supplying a pair of 3.58-MHz gated switches that produce R-Y and B-Y chroma demodulated signals, with the G-Y matrixed from the negative outputs of each. A simplified schematic of one-half of the synchronous demodulator is shown in Fig. 12-29, with one reference sine-wave gating input and one chroma receiving terminal. Each composite stage is completely balanced to remove switching voltages, and the three (RGB)-Y outputs go into emitter follower drivers for the luminance mix either in additional cathode ray tube drivers or the CRT itself. Phase angles for the 3.58-MHz inputs are tunable so that any CRT phosphor combination or "cosmetic" need may be fulfilled. With the new phosphors, phase angles from about 100 to about 105 degrees are common among the more sophisticated receivers. There are still, however, some X and Z demodulation sets that have phase angles of less than 90 degrees.

Color Killer (Fig. 12-30)

The color killer is a 6-transistor configuration, with a novel input and two pairs of difference amplifiers. But instead of operating on the chroma IF amplifier as is the usual case, the killer removes all signal from the demodulator, thus stopping all color processing.

Transistors Qa and Qb constitute a Schmitt trigger, with Qa normally off and Qb normally on. When the incoming automatic chroma control voltage increases enough to turn Qa on, current is robbed from the collector of Qa, and starved Qb shuts off. Voltage rises to almost Vcc across R2, and turns on Qc hard. The low output at Qa's collector shuts off Qd, Qf then turns on, but Qe is turned off. The DC level to the demodulator continues, but there is no chroma output because emitter follower Qe is not operating. This, Norman Doyle explains, is the reason that the "killer action does not alter the DC level at the demodulator output, thus permitting direct coupling to the video output stages."

Fig. 12-31 shows the IC with all peripheral components and connections.

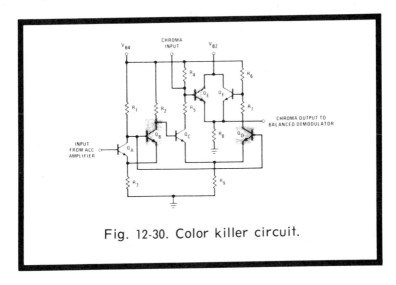

Fig. 12-30. Color killer circuit.

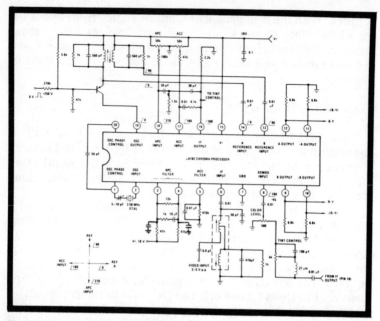

Fig. 12-31. Complete chroma circuit block diagram with all external connections. (Courtesy Fairchild Semiconductor)

QUESTIONS

1. What three inputs must **all chroma** subsystems have?

2. Where does luminance appear in the chroma circuits?

3. What function does a diode comparator perform for the subcarrier oscillator and first bandpass amplifier?

4. Is the fixed phase angle of the demodulators shifted internally or externally prior to demodulation?

5. In any graphic display of reactance, a capacitive Xc is always shown in what direction?

6. Name the three phase angles normally used in color demodulation? How are they decided upon?

7. In the early receivers, where were chroma and luminance normally matrixed?

8. How are the green colors normally derived by most demodulating systems even today?

9. What is the advantage of color CRT cathode drive?

10. Name the frequency to which all other sync frequencies in the color transmission system are slaved?

11. How much phase shift is there between I and Q and R-Y, B-Y?

12. The chroma input is limited to a 1-MHz bandpass between what two limits? What two sidebands?

13. If you have chroma cutoff only, what control would you automatically adjust first to see if it was at fault?

14. Why must color amplifiers or luminance-chroma amplifiers be cut off during retrace times?

15. How can you tell the least and most efficient phosphors of a CRT with a rainbow pattern RF input?

16. Does the Motorola MC1398 color processor have a phase comparison correction for the 3.58-MHz oscillator or an injection lock?

17. What is unusual about the MC1398's color killer?

18. In how many phases are the 3.58-MHz signals applied to the IC demodulator?

19. What does Motorola's Instamatic feature accomplish in the chroma circuits?

20. What's another trade name for Instamatic?

21. What does Instamatic control?

22. Synchronous detection can also be called synchronous s——— g.

23. To align Zenith's three-IC chroma subsystem, how many transformer adjustments are there?

24. Chroma alignment, including crosstalk adjustment, requires the use of what **two** instruments?

25. Does the color killer in the Fairchild-Zenith IC group affect both sets of chroma amplifiers?

26. Are modern CRT phosphors the same as those originally specified by the NTSC?

27. Does the uA780 have an injection lock or solid-state comparator subcarrier oscillator correction?

28. In a crystal ringing circuit, what must be observed?

29. What is a prime advantage of a crystal ringing circuit?

30. In the CA3067 chroma demodulator, complete the statement, "each pair of synchronous switches are 180 degrees———.''

31. What special circuit characterizes single chroma processor chips put together by Fairchild and Warwick?

32. What color cutoff circuit is common to both the Motorola and Fairchild chroma processor chips?

Chapter 13

Transmission Lines & Antenna Systems

Probably no subject is more overlooked than antennas and transmission lines. Now, with cable TV bringing good-to-excellent pictures into many homes and more just around the corner, it's time everyone paid more attention to what's coming from the broadcast stations.

TRANSMISSION LINES

Fig. 13-1 shows the percentage of signal remaining on new, clean lines, then on wet, dirty lines, using Belden statistics as an example. The best is also the least as you will observe in the flat twinlead chart showing 78 percent signal reception with new, dry lead, but only 1.5 percent useful signal with old, wet, and dirty lead. And think how much of this inexpensive ribbon is laying on rooftops, taped to masts, looped over gutters (which are parallel to the incoming signal polarity) or stuck to metal conduits and air-conditioning ducts in hundreds of thousands of homes through America.

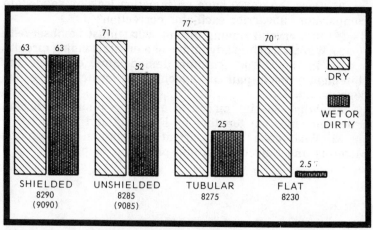

Fig. 13-1. Percentage of signal delivered to the set with clean, dry transmission line and wet, dirty cable.

Good transmission line, like anything else, usually costs money. However, many homeowners will happily pay the extra price if you explain the difference. Here's how. Belden, for instance, recommends the usual ribbon or flat lead for indoor use only, B&W VHF and FM reception only. The older tubular twinlead is all right for B&W VHF and-or perhaps local UHF reception. Coaxial cable is good for master antenna systems such as hotels, motels, apartment buildings, etc., but you'll find it lossy and somewhat capacitive over long distances. Belden recommends 8285 or 9085 (71 percent reception when dry and 52 percent wet and dirty) for all-channel systems in both color and B&W. The same cable also is excellent in fringe areas where there is little interference and only small amounts of soot, salt spray, or heavy smog. At the top of the list, though slightly signal restrictive, is Belden's shielded 8290 (9090) twinlead, especially developed for installations with signal problems such as auto ignition noise, transient pickup and excessive ghosts. Shielded lead-in may be taped to masts, routed through conduits and over metal objects. Best of all, no standoffs or lossy matching transformers are required, since the nominal impedance approaches 300 ohms. There is no signal current flow in the shield, and both 8290 (9090) and 8285 (same cable without the shield) or 9085 are specifically designed for color television.

The difference between the Belden 8290 and 9090 and the 8285 and 9085 series is that the newer "90" group is smaller in diameter than the older type, more flexible and easier to install. However, it has a slightly higher velocity of propagation characteristic and, in the case of 9085, 0.3 pf more capacitance per foot, plus somewhat more attenuation per 100 feet when the 9090 and the 8290 are compared. Also, the wire size (AWG 26) in the 9090 is smaller than the 8290 lead-in, which is listed as 22 AWG. Probably only under very extreme or unusual conditions could you possibly notice a difference in color or B&W reception.

In Fig. 13-2, we can see the percentage of signal remaining or db loss per hundred feet, which shows what happens to signals in the VHF and UHF ranges where lead-in has become brittle, dirty, wet, and thoroughly contaminated. The left vertical column is marked off in percentages, while the adjacent db figures are in parentheses. Observe that with increased frequency (VHF to UHF) the overall loss for all types of cable increases dramatically under unfavorable conditions. From this chart it is obvious that you must have 10 db or better UHF antennas where either distance or weak signals are reception factors. And a further very reasonable suggestion is

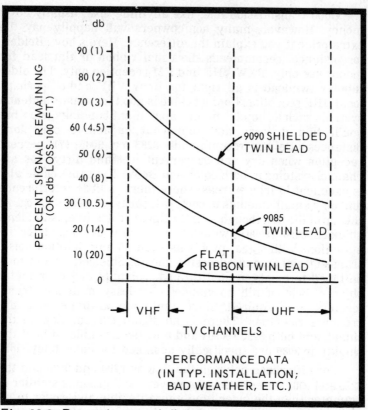

Fig. 13-2. Percentage and db losses for transmission lines when contaminated or wet.

that you inspect each antenna installation at least generally when servicing **any** receiver that indicates weak or ghosty reception. Further, the old conical or V-type antenna that did a half-way job for black and white may just kill color reception on one or more channels and produce multiple headaches unless your initial problem is recognized.

Now, you might ask, "Why is one transmission line better than another?" In any transmission line there are voltages and currents (Fig. 13-3), both coming and going, that can be calculated as the sum of the incident (outgoing) wave and the reflected (incoming) wave. If the line is an open stub, the voltage at the reflecting end is maximum and the current minimum. But if the line is shorted, the current is maximum and the voltage minimum, again at the end furtherest from the signal source. Voltage or current along any transmission line can have maximum and minimum amplitudes and the standing wave ratios (SWR), can be expressed as the ratio of either the Emax divided by Emin voltages or Imax divided by Imin currents. However, IF a transmission line is terminated in an impedance that matches the impedance of the line itself (characteristic impedance), there will be no standing wave ratio and the line will deliver as much signal as it can without interference or undue loss. At this point, the load impedance equals the characteristic impedance, and the power-factor angle of the line is zero, making it virtually a pure resistive impedance. Obviously, the closer a 300-ohm line matches a 300-ohm antenna and a 300-ohm tuner, the more signal can be delivered to any terminating load.

Between any two wires that are close together, there exists an electric field. Naturally, there is also capacitance, and a small leakage current that flows between the two wires, usually expressed as 1 divided by R equals G. This quantity is known as conductance and is calculated in 10^{-12} ohms per foot. Transmission lines also have properties of resistance, inductance and capacitance, and these are called distributed constants since they appear over the entire length of the line.

When voltage drives current in a transmission line, fields or lines of force encircle the conductors. Around one conductor is a current field (Fig. 13-4) that flows in the opposite direction to the current field set up in the other conductor, while across the insulator separating the wires is a large electric field that exerts a force on any electric charge, which could be an electron or ion. The lines of force in the electric field all move in similar directions, forward at the bottom, reverse at the top. This is illustrated by the flat lead-in drawing in Fig. 13-4. Since both fields normally operate together, they are collectively known as the lead-in's **electromagnetic field.**

Now, take a hard look at Fig. 13-4. The drawing on the left represents the flat lead that, when wet and dirty, passes only 1.5 percent signal. Notice that the electro-magnetic field completely **surrounds** the insulator and is subject to all weather and external elements. Cellular core lead is a little

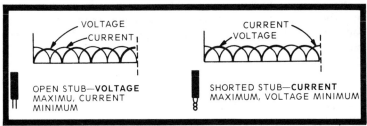

Fig. 13-3. Voltage and current on open and closed transmission lines.

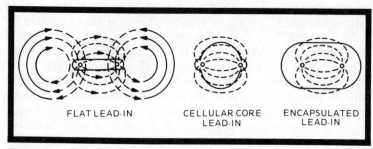

FLAT LEAD-IN   CELLULAR CORE   ENCAPSULATED
          LEAD-IN   LEAD-IN

Fig. 13-4. Electromagnetic fields surrounding three types of lead-in cable. Observe that the encapsulated dielectric contains almost the complete field, substantially reducing skin losses.

better, but some of the EM field still encircles the outside of the cable. The encapsulated lead-in, however, contains virtually all of the surrounding electro-magnetic field, has proper spacing between conductors, and should be almost impervious to all but the severest weather because of its oval, thick jacket made of a special dielectric that is an element-resistive material. If this encapsulated lead were shielded, it would contain the entire electromagnetic field and, although attenuating the incoming signal slightly, the normal ghosting and noise pickup of ordinary unshielded cable would be eliminated. Obviously, good lead-in does make an enormous difference. And don't forget to give all unshielded lead one turn per foot when connecting it to **insulating** standoffs, since it will help reduce the SWR. Shielded lead comes straight down, but its ground wire must be grounded.

ROOFTOP ANTENNA

Installing a "price" antenna is probably the surest way to deceive a customer and earn for yourself a lousy reputation. Jobber salesmen must sell their "lines" to earn a living, but it doesn't mean you should always buy. Inspect the merchandise in advance, ask questions, and request every bit of information available on the antenna. Above all, doh't buy in a hurry, or because someone says "it's good." There are such things as factual data, including gain and lobe patterns and wind resistance guarantees. Further, different antennas by the same manufacturer have all sorts of varied applications, and a study of the characteristics could save you considerable money, plus invite an excellent reputation. And this does not apply solely to those engaged in full-time antenna installations, because often the problem of poor color or ghosty pictures is directly connected to what is perched on the roof. Solid knowledge here is very often invaluable.

For all you know, some new recommended antenna may be flimsy or strong, have little or enormous wind resistance, be adequate on low-band channels and literally worthless on high channels, pick up corrosion in 48 hours, have an appalling 1:1 front-to-back ratio, be the most unsightly kluge on the block, and have all the wrong characteristics for the right neighborhood. We didn't mention characteristic impedance, standing wave ratio, UHF characteristics—if it has any—the number of directors and reflectors, or db gain.

Of course, there can be an element of reactive surprise in any installation because all factors just can't be positively predetermined. But a good, careful start usually means a happy ending. Theory says that an antenna is resonant when its measured length is half the wavelength of an incoming signal. So cutting a simple dipole for a specific frequency is not difficult. But making one antenna right for frequencies from 54 to 216 MHz and, sometimes, all the way to 890 MHz, takes a bit of doing. And everybody is not wholly successful, especially with the UHF-VHF antenna combinations, because it's a hard thing to do. The big antennas have many forward directors, often rear reflectors, and narrow beamwidths. Small antennas have wide beamwidths, few reflectors, but often side and back lobes

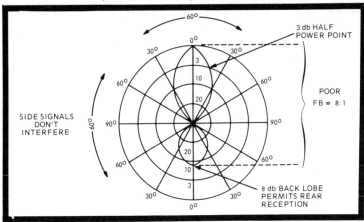

Fig. 13-5. Polar pattern diagram with front and back lobes indicating good directivity, but a poor front-to-back ratio (FB) of 8 db. (It should be at least 15 db or better.) Side lobes are nonexistent, but this antenna would pick up many signals from the back and be highly directional. Therefore, if stations were not in line, you would have to use a rotor.

that will certainly pickup stray and surrounding signals—and these can mean leading or trailing ghosts in the picture. So your only protection is to learn to read the gain and lobe patterns that **every** manufacturer should furnish with an antenna. In this way, you at least have a fighting chance for a reasonable installation at a reasonable cost, along with a satisfied customer.

INSTALLATIONS

Before considering the technical aspects of an installation, a list of DO and DON'T rules should help those unfamiliar with the basics of the task.

DO

1. Buy substantial mounting brackets and, if possible, stainless steel straps (banding) for long-lasting installations. They cost more, but stay put.

2. Use coated steel masts instead of aluminum masts for strength, especially in regions where high winds are prevalent.

3. Masts approaching 10 feet tall or taller should be guyed for added strength wherever possible. Five-foot masts usually carry almost **any** antenna **if** the mounting brackets are sturdy. Do BOLT together connecting sections.

4. Consider weather protected antennas and masts over those which are uncoated. Weather, salt spray, smog, corrosion, and dust can more easily contaminate those that aren't.

5. Check the antenna gain and lobe patterns **first** before buying any antenna. A quick analysis of the installation area, coupled with front-to-back ratios, SWR, and pattern lobes can save many a tough situation that could plague both you and the customer for months and possibly years. Even then, a narrow band antenna **won't** produce the broad response of a log periodic. So go LP if there are many stations to be received.

6. Where stations are located in diagonal or quadrature directions (90 degrees apart), a rotor is needed unless your signal bounce factor is superb. Otherwise, choose an antenna with a low front-to-back ratio if you want stations from the rear.

7. When using VHF-UHF combination antennas at distances over 20 or so miles from the transmitter, check the signal coming in directly from the antenna—on the roof if possible with a portable TV—before making the final installation. Sometimes you'll find the lead-in and especially any

non-amplifying two or four-set couplers—will cause considerable problems with UHF.

8. If you're checking an antenna on the roof with a television receiver, tilt, raise, and lower the unit for maximum gain and orient (turn) for best reception. Any peculiarities that show up in the final installation, then, will be due to the lead-in "dress," line loss, or proximity to metal conductors. In difficult situations, try different areas on the rooftop. One-man installations, with a portable TV for a helper, often can outperform a two-man job since a single individual necessarily has to be more careful.

9. In using most **coaxial** cable, don't forget the 72- to 300-ohm matching transformers and leave a small slit or drain hole in the plastic covering where it enters the house to let water out. A waterlogged piece of coax can induce all sorts of problems.

10. You can easily put two antennas on the same chimney (Fig. 13-6) if you allow enough vertical distance, say, 4 to 6 feet, to prevent inter-antenna interference.

DON'T

1. Parallel gutters, AC current lines, air-conditioning ducts, or other metal structures with unshielded twinlead; the ghosts you'll attract will all be nightmares.

2. Put cheap installations in difficult areas where there are reflections, poor signal-to-noise ratios, or great distances. Be fair with the customer as well as yourself; it's better to lose a lousy job than suffer perpetual aggravation.

3. Hurry to buy either antennas or transmission lines without knowing their specifications. "That which is cheap is often dear" in terms of personal reputation and customer satisfaction. Usually, the best is none too good.

4. Succumb to the habit of blaming the television set for your installation troubles until you've at least tried an indoor antenna to see if there's improvement in reception. Traps and attenuators can remove a great deal of interference most receivers can't reject.

5. Expect the new "mini" antennas to solve all your general reception problems. One day, such antennas may be high-gain, low-profile, and omnidirectional, but not yet. In any area with reflections, weak signal strength, and directional needs, a good standard antenna will be needed for a long time to come.

6. Put up a big antenna in high wind areas without considering its strength characteristics. If wind velocities reach 100 mph, a 75-mile-an-hour antenna installation can con-

Fig. 13-6. JFD and RCA antennas mounted on the author's chimney, one above another. The upper antenna (RCA) supplies two outlets, while the larger (JFD) delivers good signals to four. No power is used, and 11 Washington-Baltimore stations are received, most of them very well.

ceivably snap or bend at the mast, and you can't bend it back into position without mending (probably welding) the break.

7. Tape twin or oval lead—or any unshielded lead in—to masts, standoffs, gutters, or metal of any description unless you want ghosts induced by standing waves in the received pictures. Further, never lay **any** transmission line on a roof. The consequence is signal absorption (loss) and rainwater damage, leading to added reception problems. Use standoffs in all installations at all times, and **never** allow the lead-in to touch metal.

8. Neglect the use of standoffs even with shielded lead-in, since they keep the lines taut. A loose lead-in sways in the wind and could eventually break the copper stranding as well as crack the plastic shield.

9. Tell a customer the old B&W installation is always good for color. It often is, especially if there is unusually good, ghost-free reception. But take no chances when secondary signals retard images you see as ghosts, and where weak signals show a bit of snow. Ten to one, he'll need a good, new installation, complete with the usual accessories.

10. Be bashful about quoting a reasonable price. Others get it, why not you?

WORKING RESISTIVE DIVIDERS

While keyed AGC (with average video AGC combined with keyed AGC on the way) has dealt the routine divider a near mortal blow, there are still instances where standing wave ratio (SWR) line and antenna feedback voltages and currents, along with poorly devised and selected antennas, make resistive dividers quite attractive as inexpensive receiver-to-transmission line matches and signal attenuators. Obviously, such items can be bought in nice, packaged plastic cases, but that takes all the fun out of any do-it-yourself enterprise along with an impressive demonstration of skill you can exhibit before some attentive customer. With cable TV on the increase, it may be that you'll be using such devices more than you know.

For practical purposes, there are four kinds of resistive pads: L, T, pi, and H configurations, and there are certain specific things you can do with each of them and other things you cannot.

An L pad, for instance, is good for matching impedances of various descriptions with a minimum loss. It can be devised in two versions: balanced and unbalanced. Both types are shown in Fig. 13-8. The unbalanced unit can be used with coaxial cable (the shield is grounded), while the balanced divider is good for balanced transmission lines such as 300-ohm twinlead, for instance. Equations for R1 and R2 are given in the drawing for both types, with R2 over 2 substituted for R2 in the balanced version. Therefore, if you wanted to match a 72-ohm unbalanced coax to a 300-ohm receiver input:

$$R2 = (\frac{Zout}{Zin} - 1) \times R1$$

Let R1 equal 82 ohms. So:

$$R2 = (\frac{300}{75} - 1) \; 82 = 246 \text{ ohms}$$

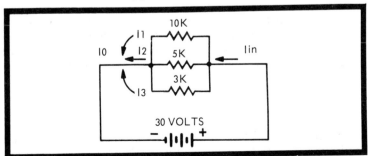

Fig. 13-7. Series-parallel to series current conversions.

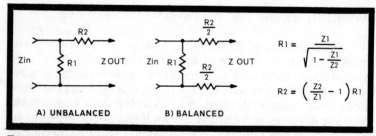

Fig. 13-8. Useful L pads in both balanced and unbalanced versions for matching different impedances with lowest loss.

or 240 ohms for round figures. If you are interested in matching a 200-ohm tuner to a 300-ohm antenna (and this could happen either way), the **balanced equation** is:

$$R1 = \sqrt{\frac{Z1}{1-\frac{Z1}{Z2}}}$$

$$R1 = \sqrt{\frac{200}{1-\frac{200}{300}}} = \frac{200}{\sqrt{1-.667}} = \frac{200}{\sqrt{.333}} = \frac{200}{.577}$$

$$R1 = 347$$

Then: $\dfrac{R2}{2} = \left(\dfrac{Z2}{Z1 -1} \right) R1 = \left(\dfrac{300}{200 - 1} \right) .347 \times 10^3$

$$\frac{R2}{2} = (1.5 - 1) .347 \times 10^3 = .174 \times 10^3 = \frac{174 \ ohms}{2}$$

So each R2 is 87 ohms—in a balanced pad, of course. And this should be a good introduction to balanced and unbalanced L pads.

You can have lots of combinations, and you can produce a pretty good match with non-phase shifting carbon film or composition resistors, Tolerances are not **too critical.** Remember, however, these L pads are **for matching different impedances** and are called asymmetrical networks since the input and output impedances are **not** equal. Now, if you were matching an unbalanced asymmetrical network, say from 50 to 300 ohms, R1 would be 56 ohms and R2, 270 ohms, while in a balanced asymmetrical network R1 would be the same 56-ohm value, but R2 would be halved and amount to approximately 120 or 140 ohms.

T and H pads (Fig. 13-9) are similar to the L pads, in that the T pads are unbalanced while the H pads are balanced. However, these two (as opposed to the L pads) are symmetrical since they are signal attenuators situated between matched impedances and do nothing but absorb some of the voltage and current between antenna and receiver. Handling the equations for these two networks is just a bit more tricky than the L pads, since you're using a signal attenuation ratio in voltage rather than simply a straight resistive calculation. When you're talking about a 10 or 20 db loss, you'll usually have to go to a slide rule or a table of db current and voltage ratio loss figures and dig them out. We'll provide you with some of the more common ones in case these tables aren't readily available:

A 10 db loss is a current-voltage loss ratio of 0.316.

A 20 db loss is a current-voltage loss ratio of 0.100.

A 25 db loss is a current-voltage loss ratio of 0.056.

A 30 db loss is a current-voltage loss ratio of 0.032.

Voltage-current in decibels is $20 \log_{10}$ E2 over E1.

As an example, let's calculate R1 and R2 for a typical T pad that must attenuate the incoming signal by 10 db:

$$R1 = Zo \left(\frac{A-1}{A+1}\right)$$

Now, since the T pad is a symmetrical unbalanced network, we'll be using it either with 50- or 75-ohm impedances where the shield side of the cable is common ground. Let's select 75 ohms.

$$R1 - 75 \left(\frac{0.316 - 1}{0.316 + 1}\right) = \left(\frac{0.684}{1.316}\right)$$

$$75 = 75 (0.519) = 39 \text{ ohms}$$

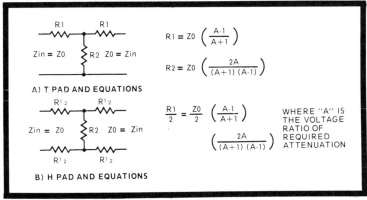

A) T PAD AND EQUATIONS

$$R1 = Z0 \left(\frac{A-1}{A+1}\right)$$

$$R2 = Z0 \left(\frac{2A}{(A+1)(A-1)}\right)$$

$$\frac{R1}{2} = \frac{Z0}{2} \left(\frac{A-1}{A+1}\right)$$

$$\left(\frac{2A}{(A+1)(A-1)}\right)$$

WHERE "A" IS THE VOLTAGE RATIO OF REQUIRED ATTENUATION

B) H PAD AND EQUATIONS

Fig. 13-9. Signal attenuating T and H pads where input and output impedances match. T pad (A) and equations; H pad (B) and equations.

$$R2 = Zo \frac{2A}{(A+1)\ (A-1)} = \frac{2 \times 0.316}{(0.316 + 1)\ (0.316 - 1)}$$

$$= 75(\frac{0.632}{0.9}) = 52.7 \text{ ohms}$$

So a 10 db pad has for its T network values, R1, 39 ohms, and R2, 52.7 ohms. The nearest commercial values would be about 33 and 51 ohms. Now, how about a 20 db padder for the H network where the input and output impedances are typically 300 ohms? This, of course, is a balanced symmetrical network and is useful in attenuating signals coming down the usual 300-ohm TV transmission line. Here's how this one is done, since you can see that a 20 db voltage-current loss amounts to a ratio of 0.100.

$$\frac{R1}{2} = \frac{Zo}{2(A-1)\ (A+1)} = \frac{\frac{300}{2(0.100 - 1)}}{(0.100 + 1)}$$

$$= \frac{150\ (0.9)}{(1.1)} = 150\ (.817) = 122.6 \text{ ohms}$$

$$R2 = Zo\ (\frac{2A}{(A+1)\ (A-1)}) = \frac{\frac{300}{2(0.1)}}{(1 + 0.1)\ (1 - 0.1)}$$

$$R2 = \frac{300(0.2)}{(1.1 \times 0.9)} = \frac{60}{.99} = 60.6 \text{ ohms}$$

And here you can juggle the values slightly so that R1 comes out to about 120 ohms and R2 expands to 68 ohms for typical commercial values. However, all R1s must be the same value. From these examples you should be able to calculate virtually any attenuation needed in any of these T and H pads.

There are also pi padder networks that can be used readily in place of T padders if you wish. The converting equations are rather easily used. This time, however, we'll adopt the term impedance instead of resistance and so make the application broader in the sense that not only will it cover resistances but reactances as well. Therefore, the symbols in Fig. 13-10 will be in impedance (Z) terminology for both the T and pi section padders. To convert from a pi to a T network, use the following equalities:

$$Z1 = \frac{ZaZb}{Za + Zb + Zc}$$

$$Z2 = \frac{ZbZc}{Za + Zb + Zc}$$

$$Z3 = \frac{ZaZc}{Za + Zb + Zc}$$

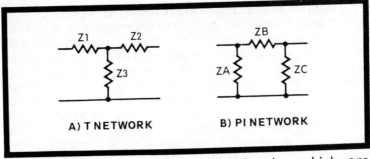

To convert from a T network to a pi network:

$$Za = \frac{(Z1Z2 + Z2Z3 + Z1Z3)}{Z2}$$

$$Zb = \frac{(Z1Z2 + Z2Z3 + Z1Z3)}{Z3}$$

$$Zc = \frac{(Z1Z2 + Z2Z3 + Z1Z3)}{Z1}$$

INTERFERENCE, GHOSTS, AND SUCKOUTS

Useful quarter-wave and half-wave elimination filters can be made from ordinary twinleads. In recent years, such filters have not been needed, generally, with the more expensive B&W U.S. type receivers, since front-end filters and trapping effects have been highly effective against CB, commercial, and other forms of high- and low-band interference. With the less expensive sets, though, you may find an occasional instance where a simple trap will make the difference between artistic success and technical failure. Especially did we find this true recently in an apartment installation where the house amplifier wasn't the best and the installation of a simple trap restored lots of gratifying color. Had the coax been a twinlead line, we might have used a piece of metal foil or our thumb and forefingers to find a point of signal suckout due to poor line-receiver termination. At that juncture we could have installed a small 2- to 20-pf ceramic trimmer across the line and tuned it with a plastic or fibre wand for best signal strength. Or, we could have left the metal piece in place, secured with good, sticky tape.

Ghosts (Fig. 13-11) can be a nuisance and sometimes very difficult to deal with. There are the usual trailing ghosts to the right of the main image where the prime signal travels a shorter distance than the reflected information and the ghost is delayed by a microseconds interval and appears after the usual picture information. You may also have a leading ghost on the left of the initial image when the receiver is quite close to a TV transmitter. This is due to either internal signal pickup, a very long transmission line from receiver to antenna and perhaps parallel to the signal polarity, or some sort of tuner trouble that used to crop up more frequently than it does now because of overheated vacuum tube load resistors when there were tube short circuits. Then there are even multiple ghosts or vertical banding troubles due, again, to tuner problems, several reflected images, or open or wrong value capacitors in the horizontal windings of the deflection yoke. With today's "hot" front ends (transistorized), you'll find antenna and lead-in problems much more prevalent than internal receiver faults. Therefore, in poor signal neighborhoods where ghosts abound, try a highly directional antenna, perhaps with rotor, and a fully shielded twinlead line. Reduce as many external possibilities as you can, then experiment further with antenna-receiver line matching.

In an exceptionally tough situation, it can pay you to actually calculate, even in a somewhat rough fashion, the **actual distance the reflected signal is bouncing** from the offending building or tower toward the set. You simply do it this way: measure in inches the width of the picture, and be sure it is not unduly overscanned; then measure the distance between some identical point in the ghost and the original picture for a ratio of, say, 1.5 inches to 20 inches; then, knowing that television signals travel about 1,000 feet per microsecond in the air and that the receiver's CRT beam is swept through one

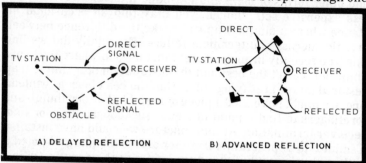

Fig. 13-11. These drawings show how trailing and leading secondary images (ghosts) are formed by single and multiple reflections. (A) delayed reflection; (B) advanced reflection.

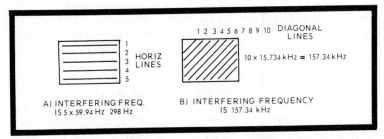

A) INTERFERING FREQ.
IS 5 x 59.94 Hz 298 Hz

B) INTERFERING FREQUENCY
IS 157.34 kHz

Fig. 13-12. How to tell difference frequency of interfering signals in multiples of vertical and horizontal sync rates. Notice that we said "difference" frequencies and not fundamentals. The interfering frequencies is 5 X 39.94 Hz. (A) At (B) the interfering frequency is 157.34 kHz.

horizontal line in 52.4 microseconds, the total horizontal sweep time amounts to 52,400 feet-microseconds, or about 10 miles. Therefore, in this instance 1.5 divided by 20 x 52,400 equals about 3930 feet, the air distance the reflecting image travels— not necessarily the actual distance between the receiver and the reflecting object. For instance, a building half the distance of 3930 feet behind the receiver could be the culprit just as easily as one at a 45-degree forward angle from the receiver. However, should there be an out-standing tall object at some known point within the approximate distance, it could well be your target. If you want to locate the direction of the ghost accurately, you can probably find it with a field strength meter and a receiver with an antenna you can move. At some particular channel frequency you will find a strong signal, then one that is not so strong; the latter should be your quarry.

Occasionally, you will have to determine if a secondary image, such as a ghost, is originating in the transmission line or is simply an airpath signal such as we calculated. You can do this by estimating the length of line, say 100 feet, multiplying it by 2 and dividing by some constant between 0 and 1, usually 0.83. If the ghost has an airpath of a thousand feet or more, the transmission line is not at fault since 100 x 2 divided by 0.83 equals 200 divided by 0.83 equals 240 feet. And, of course, the airpath of the transmission line is probably less than one fourth of a thousand or more feet.

One peculiarity you can always calculate with ballpark accuracy is the gross frequency of overriding interfering signals (Fig. 13-12). If they are horizontal, count the number of dark lines and multiply by 59.94 Hz, since this is the vertical repetition rate, and you will have the difference frequency of the interference. Or, if the lines are either diagonal or vertical, count them and multiply by the horizontal rate of 15,734 Hz. If there are too many black slant bars to count, total the number of lines in an inch, then multiply again by 15,734 kHz for a final figure.

This all sounds very easy and good, but there's just one small catch—you've been calculating the difference and not the fundamental frequency. These vertical and horizontal lines are, in truth, fundamentals which, unfortunately, you seldom find in real life. Usually, interference is the sum and difference signals between the interfering frequency and the video carrier if the TVI is 4 MHz or less. If more than 4 MHz, they are the difference signals only. So an accurate determination of a very high frequency is rather tricky. About the best way to discover your trouble is to get some sort of listing of transmitters in the immediate area, take a signal generator along and try to match those frequencies so you can arrive at the fundamental. Otherwise, you may have an aimless hunt trying to establish the interfering problem since these stray energies seldom lie still and allow you to establish a solid "fix" on their specific characteristics.

If you have an old or very cheap receiver, the addition of a high-pass filter at the **tuner input terminals** may help. However, all the newer receivers, especially the quality models, have combined IF and high-pass filters already built in; so an additional series filter would not be very beneficial, unless the interference is extremely marginal, and then only because the filter supplies an extension of what was already there.

The second point about filters is this: in 300-ohm input receivers there is already a 300- to 75-ohm transformer (called a balun) at the antenna input to the tuner. If you can imagine a 150-ohm transmission line of infinite length with the input in series to "see" 300 ohms and the output in parallel to "see" 75 ohms, you would have immediate attenuation from DC to about 5 MHz anyway, because one side of the output is already connected to ground. Naturally, the line isn't infinite, so you put in a ferrite that makes the balun look like this theoretical line and the tuner input "see" its proper impedance.

The shielded 75-ohm input, however, goes directly into the usual high-pass filter between the RF amplifier and the antenna. This high-pass filter often has 80 db rejection (60 db at any rate) for the receiver's IF return characteristics, and does pretty much the same for any incoming frequencies at and below 54 MHz as well. In some of the top receivers, there is also a tunable FM notch rejection filter to help with in-

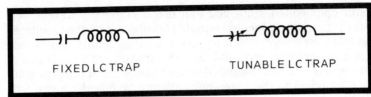

FIXED LC TRAP   TUNABLE LC TRAP

Fig. 13-13. Fixed and tunable LC traps can be used to shunt aside an unwanted frequency. A 20- to 100-pf capacitor is usually used for C.

terference from the 88-108 MHz FM-stereo band. So, really, regardless of what the "old hands" are saying about the addition of multi high-pass filters, you're probably going to have to be pretty specific with any strong interfering frequency, since it may just be overriding the input traps and any additional high-pass filter installed also.

Before leaving this topic there are two more points we should make: Normally, the interfering frequency will have to be within 500 kHz of either the chroma energy or picture carrier if you are to see a low power interfering signal. The second point is that a 75-ohm input is quite useful on a "cold" chassis since it can be grounded. On a "hot" chassis, you can't take such a ground to the antenna input terminals.

In future CATV receivers, there will be better limited passband tuners, greater shielding, and more attention paid to the video IFs where interference can be picked up also as we discussed. Reputedly, most of the forthcoming crop will double in brass—be good for air signals as well as cable.

TRAPS

Traps are used when low-pass and high-pass filters are ineffective. This condition often occurs when the receiver is in the midst of a strong RF signal area. Image frequency (the local oscillator signal plus or minus the IF; the opposite of the IF) interference may respond either to a fixed or tunable trap (Fig. 13-13). It may be reduced or eliminated by 300-ohm or 72-ohm open-ended stubs Fig. 13-14, depending on the impedance of the transmission line. You can try the easy way. Use a pair of cutters and begin chopping at the tail end of about 35 inches of line until the stub is tuned to the desired frequency, or you can use an equation that will drop you directly into the ballpark.

The length of a quarter-wave, open-ended trap for 300 ohms can be found by using the following equation: Length (in inches) equals 2450 divided by the frequency in MHz. For a 72-

ohm open-ended stub, you can use a slightly different constant: Length (in inches) equals 1945 divided by the frequency in MHz. However, you must use two equal length pieces, one for each antenna terminal, and the shields must be joined at several points and grounded to the chassis. Both types of traps are attached to the receiver's antenna terminals, along with the lead-in. The length for a disturbing frequency of 100 MHz is: 300-ohm length equals 2450 divided by 100 MHz, or 24.5 feet.

In addition to quarter-wave stubs, there are also half-wave stubs. Over the years we have found half-wave stubs easier to use, since fewer reflections or ghosts are usually induced with this type of shorted stub. The difference between quarter-wave and half-wave traps is basically the terminations. In the first, the voltage is maximum at the open end while the current is minimum. In the second, the current is maximum and the voltage minimum at the point of termination. The equations are also somewhat different. To calculate a half-wave 300-ohm stub: Length equals 4,850 divided by the frequency in MHz. For a half-wave 72-ohm stub: Length equals 3,900 divided by the frequency in MHz.

You must, however, recall one very important point: A quarter-wave stub also attenuates second harmonics, while a half-wave stub attenuates third harmonics, in addition, of course, to the fundamental. So a 90-MHz signal attenuated by a quarter-wave trap would also give you problems on Channel 8, between 180 and 186 MHz. A half-wave stub would attenuate frequencies at 270 MHz also, if cut to the same 90 MHz fundamental. So when using traps, be careful you don't interfere with a signal, and do check all receiver channels before putting the job to bed.

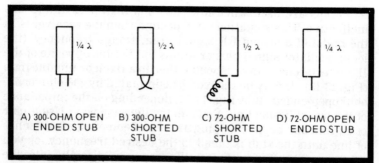

Fig. 13-14. Different types of stubs may be used to bypass and reject interfering frequencies. (A) 300-ohm open-ended stub; (B) 300-ohm shorted stub; (C) 75-ohm shorted stub (D) 75-ohm open ended stub.

QUESTIONS

1. When you use ordinary flat twinlead in dirty, foul weather, what percentage of signal do you expect it to pass?

2. What cable should you use in ghosty and noisy areas?

3. As frequencies increase, do cable transmissions become better?

4. What's a good practice before servicing **any** receiver with an outside antenna?

5. Why is one transmission line superior to another?

6. When do you decide on an antenna for any installation?

7. A narrow front pattern lobe is good for directivity. A good front-to-back ratio is often desirable. Is there a long distance antenna with a broad beamwidth and no side lobes?

8. Can you mount two antennas on the same chimney?

9. Why bolt together 5-ft mast sections?

10. Why install stainless steel chimney straps?

11. Can you lay ordinary twinlead on a roof? Why not?

12. Where do you find leading edge and trailing edge ghosts?

13. In ghosty neighborhoods, what do you do?

14. 75-ohm inputs in the new receivers often have a high-pass filter input with a —————— to —————— db rejection for IF reflections.

Chapter 14

Troubleshooting

With solid-state and modular receivers already on the market, a very considerable change will have to be made in the selection and shop use of test equipment to service these new products efficiently. With the low operating voltages (less than 1 volt) encountered in some semiconductor circuits, the gross inaccuracies of previous amateur gear, such as inexpensive meter kits and $150 oscilloscopes, are totally inadequate. Also, in such receivers you will find a growing number of 2 and 5 percent resistors, 5 and 10 percent capacitors, fine tolerance coils, and many closely regulated power supplies. If your service equipment can't be calibrated closer than these 5 percent average allowable circuit component variations, then many measurements will be highly inaccurate or completely worthless, and you won't be able to make the needed repairs.

True, module replacement will satisfy over 80 percent of the problems in the new receivers, but the other 20 percent (chassis breakdowns) will be extraordinarily difficult if your equipment isn't right. At the local distributor's, we're noticing more and more solid-state sets that have been badly "shotgunned" by service people who either don't know how or refuse to use adequate equipment for almost any sort of repairs. The television business is a good business only if all test equipment is both relatively new and good, and servicemen are thoroughly trained in its use. Transistor and integrated circuit receivers simply require new and more sophisticated methods of attack with very modern waveform display devices and signal generators.

DC AMPLIFIER, TRIGGERED SWEEP OSCILLOSCOPE

During this discussion, we progress from the exotic to what might be termed "standard" service equipment to demonstrate what really can be done with the best, then with the above average but absolutely not second rate gear.

To use an oscilloscope effectively, you must exploit all of its advantages with considerable efficiency and be competent in interpreting what you see. Initially, any oscilloscope you use should have a 10-millivolt deflection factor, a bandpass of at least 10 MHz (plus or minus 3 db down) a calibration accuracy of no less than 5 percent (with 3 percent preferred), and DC vertical amplifiers that are linear throughout from 10 millivolts to 20 volts or more (0.1 volt to 200 volts per division with a low-capacitance 10X probe). A dual-trace scope is preferred, but a single-trace model is acceptable, except servicing more than one section of a receiver at one time will be somewhat more difficult. A scope should have sufficient high voltage for full, fast trace resolution, some power supply regulation, calibration ranges that are easily set, and a horizontal sweep section and amplifier that will sweep from at least 0.5 seconds to 500 nanoseconds, with 100 to 200 nanoseconds favored. Those are the basic requirements, but you can be fancier if your pocketbook allows.

What's all this cost? Prices will range from $400 for a single-trace scope to $500, $600, $975 and up for the better dual-trace units. The one we use first happens to be an exceedingly fine Hewlett-Packard Type 180 (Fig. 14-1) with 1801A dual-trace 50-MHz amplifier and 1821A time-base generator, a package that's worth $2390. Obviously, we don't expect service people to buy "Henry's Mark 4," but something in between a great oscilloscope and an inaccurate AC-coupled excuse is most necessary and desirable. The Tektronix 7403N dual-trace scope, which we use later, is shown in Fig. 14-2.

Any serviceable oscilloscope has three outstanding uses: peak-to-peak voltage readings for AC measurements, DC readings for voltmeter measurements, and time-base readings for time and frequency measurements. Anything less, say AC only, actully is only one-third of an oscilloscope, and such basics are hardly worthwhile. Of course, the RMS (root mean square) work ability of a sine wave is equivalent to a DC voltage. So if your oscilloscope is calibrated in DC volts, for instance, the RMS voltage X 1.414 equals peak and X2 or RMS X 2.828 equals peak-to-peak.

Pulses and square waves go from one DC level to another, so the oscilloscope can follow these transitions also for accurate readings. DC-wise, you simply set the amplifier trace at some point on the graticule, and the number of semi- or complete divisions the trace rises or falls is the positive or negative DC voltage, respectively. If you are looking at an AC waveform and want to know on what DC voltage it's "riding" or "hanging," simply flip the DC-AC selector switch, and the number divisions the AC trace goes up or down times the vertical attenuator setting tells you the DC level. When a trace

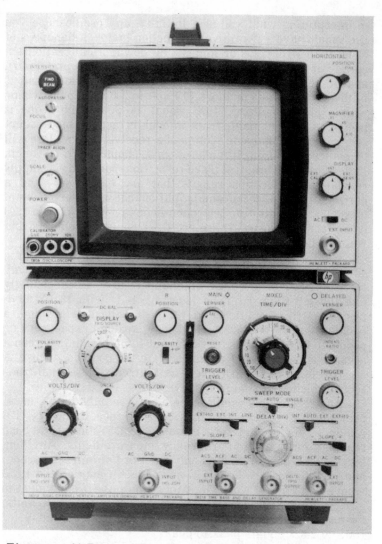

Fig. 14-1. H-P's 50-MHz dual-trace scope with A and B time bases, mixed sweep, and a time base that (with 10X magnifier), will display waveforms at 10 nanoseconds per division.

(amplifier) is reading a pure AC input, the waveform is half above and half below DC reference level 0, or "ground."

To demonstrate how an oscilloscope can be used to find trouble in tricky circuits, let's take the scope into a horizontal sweep system.

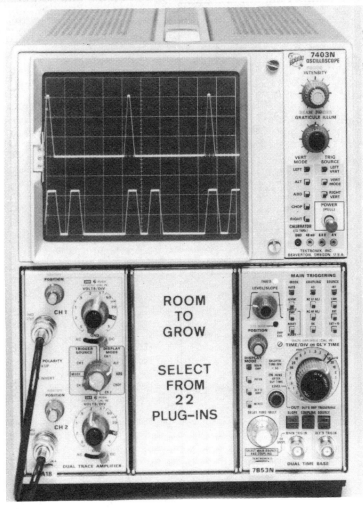

ig. 14-2. Tektronix 50-MHz oscilloscope with three plug-in nits. It is accurate to 2 percent, with 1.22 cm per division. he unit features an 8 x 10 graticule, 5 mv per division ertical sensitivity, two time bases, and room to grow.

Cantankerous Case of Horizontal Flops

In the RCA CTC46, RCA's instructions for oscillator ignment are to bias off the incoming sync signal to the phase litter, disrupt an RC time-constant to the sine-wave coil, ljust the horizontal hold for a slowly or non-floating picture,

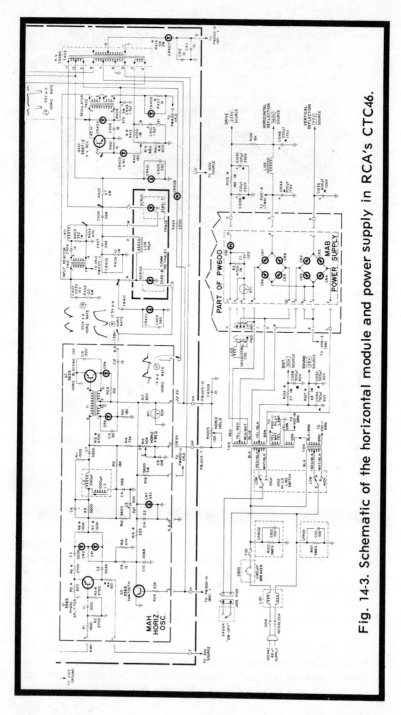

Fig. 14-3. Schematic of the horizontal module and power supply in RCA's CTC46.

then do the same for the sine-wave coil, removing one jumper as each assignment is completed. That done, you hopefully turn on this completely IC and transistorized receiver, expecting the picture to be neatly framed. But, when this set was turned on (with antenna connected) the picture was somewhat out of horizontal sync, but in about 30 seconds corrected itself.

With a color-bar generator connected to the antenna terminals, the set was turned back on after a short turn off period. There was a distinct sync bend (possibly 60 Hz) that didn't go away for about 30 seconds. Now, theoretically, you should have one of two conditions: either the thermistor in the degaussing circuit (Fig. 14-3) is faulty or there is a bad filter capacitor in a 60-Hz power supply line. If you go back to the MAH horizontal oscillator, you'll see what appears to be a flyback pulse feed to the blocking oscillator through CR408, but that isn't the real circuit at all. You'll also discover that the 22.5 volts shown on the schematic turns out to be nearer 33 volts, since there is a zener regulator (CR7) for this entire line connected from one end of R17, with a 50K thermistor to ground from the other end and this won't produce much of a voltage drop. CR408 turns out to be an AC blocking diode and is the pathway for the high-voltage regulator collector supply. If you measure across 5-watt R409, you'll find only a difference of 48 to 56 volts, which is a bit of a startler when you're expecting something else. But, indeed, it's perfectly all right. Right away you begin to question the presence of any 60-Hz stray hum in the circuit—and there can't be any, since both the 160-volt and 30-volt sources originate from bridge diode rectifiers and these just don't produce 60-Hz ripple **unless** one of them is defective. However, if that were the case, you wouldn't have 160 volts varying slightly with signal because of non-regulation and a 30-volt source that does the same, but with less magnitude because of the lower voltage and additional filtering and, therefore, more regulation.

If we jump to the top of zener regulator diode CR7 and find 33 or 34 volts there, the entire module is getting all its rightful voltages. After this, we can profitably examine both the video and horizontal oscillator waveshapes at the junction of C7, R11, and L2, the output portion of the sine-wave coil we have always tuned so that all peaks are on an even plane. In Fig. 14-4, the top trace was picked up at the sync separator. We increased the scan rate of the scope to 10 microseconds per division (Fig. 14-5), and the horizontal oscillator waveform and the video appeared spread out. The waveforms in Fig. 14-5 are something exceeding one line of video information and a little more than one cycle of horizontal oscillation—both

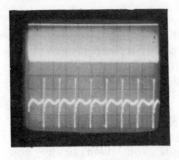

Fig. 14-4. Horizontal oscillator sine-wave outputs with all tips equal.

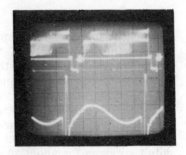

Fig. 14-5. Video and sync at 10 microseconds per division. Top amplitude 2v per division; the bottom is 0.5v per division.

relatively normal, except the difference in peaks of the oscillator transition amounted to 0.5v at one half volt per division. But since we had just retuned the entire oscillator according to RCA's instructions, we let it go temporarily. The video looked good and the burst on the sync pulse back porch is outstanding. The small overshoot at the trailing edge of the horizontal sync pulse can be ignored since it is in the no-picture period anyway and could not be seen.

For curiosity (Fig. 14-6), we turned to another station, using the same scope time interval, but not the same amplitude on the bottom trace (1v per division), and found the horizontal sync pedestal more uniform but the 3.58-MHz burst was considerably lower in amplitude. Yet, strangely enough, if you went to the broadcasting station, they would probably have the magnitude of this "varying" color carrier exactly right. And this simply means that in your local area, you'll have to learn (on a good receiver) the nominal signal inputs from the antenna and what to expect in the way of normal or seemingly abnormal transmissions. In Fig. 14-5, the top trace is set at 2 volts per division, with the zero reference at the center horizontal graticule line. The bottom trace is calibrated for 0.5 volts p-p, with the zero reference on the next line to the bottom.

A picture of one almost complete horizontal line sweep interval (at 10 microseconds per division) is seen at the top in Fig. 14-7 (10v per division). The sine wave (bottom, 1v per division) was set for equal peaks, just for curiosity. The 30-volt horizontal timing pulse was all right, but the sine-wave coil tuned for symmetrical peaks just didn't work out. So we retuned for the unsymmetrical waveform and took a look at the horizontal sync pulse through the differentiating network and at the base of the phase splitter transistor, Fig. 14-8. It's total amplitude was 15 volts p-p, and the 45-microsecond straight line portion was just about at DC. So certainly the amplitude is sufficient, even though the trailing edge of the pulse (at 10 microseconds per division) has an initial fall time of almost 10 microseconds before its down time becomes steeper and the total fall width is only about 5 microseconds. Suppose it had been 10 or more? Wouldn't this have caused all sorts of sync problems? It certainly would have.

What about the power supplies? Do they come on instantly when the receiver does? Any unusual ripple or transients getting through? As previously checked, the DC values were entirely adequate, and both 160-volt and 30-volt supplies appeared immediately when the on switch was engaged. What about ripple? The top trace in 14-9 was set for 0.5 per division AC only, and the bottom at 0.2v per division AC only. So in the top trace you have some 0.7v ripple and bottom about 0.2v ripple with, of course, some 1.5-volt SCR high-voltage spikes that don't show up on the other sides of the decoupling resistors, such as 5-watt R18 (the 160-volt input) and R4, the 30-volt input. It's also possible that the SCR spikes were picked

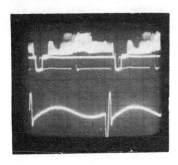

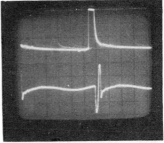

Fig. 14-6. Same signals as in Fig. 14-5, but the waveform is calibrated at 2 volts, while the bottom is 1 volt. The receiver was tuned to another station.

Fig. 14-7. The top trace is the horizontal sync pulse; the lower trace is the sine-wave horizontal oscillator output, 10 volts per division and 1 volt per division, respectively. Time base, 10 microseconds per division.

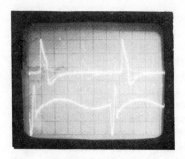

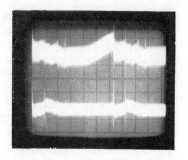

Fig. 14-8. Waveform at the top (5v per division) was observed at the base of the phase splitter. Jitter is normal for AFC diodes and this type of circuit, since frequency searching is continuous.

Fig. 14-9. Normal ripple of 0.5v per division in the 160v supply and 0.2v per division in the 30v supply; top and bottom traces, respectively.

up by the probes and not in the power supplies at all, but this isn't altogether plausible since there's no outstanding trace of the spiking problem once you're into the horizontal circuit. We did find, however, that the voltage at point HH was much nearer 30 volts than the 22.5 listed on the schematic—and as it should be because of the 33-volt zener (CR7) regulator. R424 is 3.3K and precedes CR408, being part of a time constant with filter C409.

How do we know the waveform spikes in Fig. 14-9, which are uniform distances apart (one in each cycle), are from the high voltage? Look at Fig. 14-10. The top trace is the horizontal oscillator drive to the SCR at 2 volts per division, point G. The bottom trace, at 100v per division, is the retrace SCR waveform. These traces are synced on alternate sweep cycles within the scope. You can see the small pip at about 15 microseconds of the conduction period (top) coincide with the trailing edge of the SCR waveform cutoff. In dealing with NPN pulse circuits, of course, the conduction trace is near or at 0 when the transistor is conducting, and at the power supply level when off. So here, the drive pulse is 6 volts and the retrace SCR goes for just over 300 volts p-p, with a conduction time of just about 25 microseconds, since the scope time base is set for 10 microseconds per division.

Now, what if the filter (C4008) at the end of the horizontal hold is defective? C4008 is connected to the forward-opening

Fig. 14-10. The oscillator drive waveform (top, 2v per division) shows a small spike trigger and retrace SCR output with small spike showing in the oscillator forward trace one division past the center of the graticule.

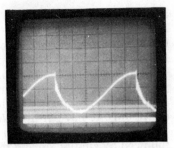

Fig. 14-11. These scope traces show the charge (faint trace) and discharge of C4008. The 1.5v trace is the residual waveform.

control panel, a rather formidable task, so it seems, to get to. So use an oscilloscope to check the capacitor. In a DC condition such as this, the job should be rather easy. The nearest path to ground is through the 3.3K resistor, diode, and HV transformer, or through the blocking oscillator, itself, if it switches completely into saturation, something we don't (at the moment) know. But, say, we have a theoretical 50K resistance for this capacitor to discharge through. The time constant would be $0.01 \times 10^{-6} \times 0.5 \times 10^{-6}$ equals 5×10^{-3} or 5 milliseconds. Now, it's true we can't tell if there is slight leakage, and there may be. But we can certainly tell if the capacitor is open or shorted, and here's how: Since we're dealing with a 30-volt DC swing, set the oscilloscope's vertical amplifier (using only one channel this time) to 5 volts per division (Fig. 14-11) so that the trace actually covers six vertical divisions, and it should do so in the time of 5 milliseconds or less. Turn the receiver on and see what happens. In this instance, the trace bounced up very fast. If you look sharply, you can see a slight and light wave reaching to about the sixth horizontal graticule division from the bottom.

The next test is to see if this capacitor discharges at about the same rate. When the receiver is turned off, the two "traces" near the bottom are actually the bounce and settled traces, with the more intense bottom band about 2.5 volts above zero reference, which is the bottom horizontal graticule line. So the capacitor doesn't discharge completely when the power supply is cut, but it does lose its charge rapidly enough to be evaluated as probably good.

What should this residual waveform look like? To show you (still Fig. 14-11, upper trace), we did another bit of trick photography, taking a double exposure of the trace amplified on AC only at 0.5v per division. As you see, it's a 1.5-volt trace with a sharp positive pip for at least a simulated sync pulse, followed by ringing during the downward conduction time and a smooth trace as the waveform ascends toward cutoff. Shunting another capacitor across terminal HH to ground did reduce the amplitude of the waveform but seemed to increase the amount of the ringing. So we must presume the other inductance and capacitances in the circuit, including C4008, are possible ringers, too.

If you really want to check leakage here, just pull the MAH module and put an ohmmeter from HH to ground. In the open horizontal hold position, the measurement across both the control and to ground is 9K. In the closed (zero) position of the hold control, the measurement across the capacitor to ground is still 9K. R4205 should look almost like an open circuit, since it isn't connected to anything but CR408 and R424 plus the HV transformer. Let's make one final check. With the hold control closed (zero resistance) the measurement from terminal HP on PW400 measures 15K. Yet, there's the diode and resistor still connected to ground through the secondary of the high-voltage transformer, and this is the other 9K.

In desperation you could pull the control assembly from the back side of the chassis (it's held together by two ¼-inch screws) and change this suspected capacitor; it may or may not do any good. The one in the circuit showed a slightly excessive power factor. For curiosity, we took the assembly and its connecting cables apart and changed the capacitor just to go through the process and to look at the quality of the part. After finding the excessive power factor (dissipation the capacitor shouldn't have) the part was replaced, but with no real certainty this was a cure for the problem simply because the fault seemed too minor.

With the new capacitor installed, we decided to make two more checks: In Fig. 14-12, you see the blanking pulse (top) from the feedback loop on the transformer at 10v per division, at terminal H a total of some 30 volts, and this checks. Below it is the sawtooth output of Q3 at 5 volts per division. Both pulses are acceptable since they reach the proper amplitude; neither is on during the 11-microsecond horizontal blanking interval

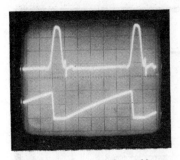

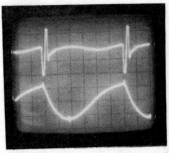

Fig. 14-12. High-voltage blanking pulse (top waveform) and sawtooth drive (lower waveform). Notice that the drive "bottoms" no longer than the period of blanking which is 10 microseconds.

Fig. 14-13. Peaks are even in the oscillator output (top), and the lower drive pulses are clean. But this isn't the solution, believe it or not.

as you can see by the positive pulse off time of the upper trace and the Q3 cutoff before the linear rise of the bottom trace. Now, what about the linearity of the trace at the junction of the horizontal hold and our new C4008, after, that is, we readjusted the sine-wave coil? Fig. 14-13 shows the results. There is no ringing in the lower trace, and the upper sine-wave peaks are exactly even at the maximum points, touching the second horizontal graticule line from the top. Another interesting thing: The horizontal hold, if moved steadily across its total range, does not disturb the oscillator lock in the slightest. If moved fast and clicked at each end, however, it will throw the oscillator out of sync with about four or five broad horizontal bands. Are we really repaired? Well, let's let this "intermittent" be turned on and off at several long intervals for a while and we'll be positive.

Sometimes, such careful waveform analysis is entirely necessary. And when you're doing it, you have to analyze the circuit operation as well. It's also emphasizes the fact that many of these modular plug-in boards can be serviced while plugged into the receiver. It's simple to insert a substitute, but normal circuit problems are no more difficult than some much advertised "hand-wired" circuits. Actually, as far as the author is concerned, plug boards are easier to troubleshoot than the old point-to-point terminal strip connections. The big difference is that you just don't pull components off the boards and replace them at random. Many are specialized types, as you will discover by consulting the parts list, and others are unique enough so that only the manufacturer is your supplier.

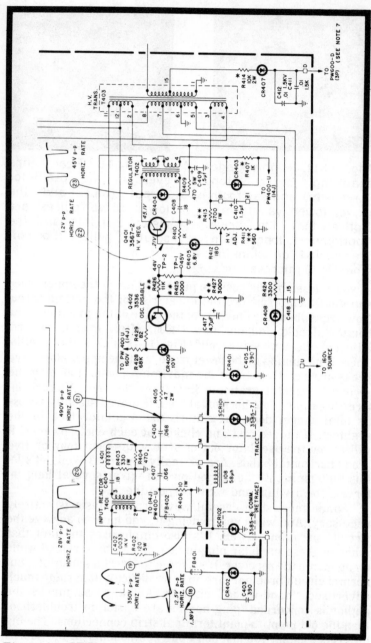

Fig. 14-14. CTC54 partial schematic with oscillator disable change included. This will not be included on your CTC46 schematics since it's a recent FCC requirement.

Therefore, in the future, if the fault isn't on the board, try an analysis such as the CTC46 problem just demonstrated, and with a little practice many joyous things should come to pass.

We deliberately picked a problem that could have been in the AGC, sync separator, horizontal oscillator, a bad return winding on the flyback transformer, defective AFC diodes, or a few other passive components. Seldom do you collide with such stubborn, unyielding difficulties, since complaints ordinarily are directed at permanent breakdowns. But once in a while this is the case, and then you are stuck with the same sort of investigation just undertaken.

Meanwhile, we did find that resistor R10 had been changed from 3.9K to 1.2K, for some reason, probably to speed up the anti-hunt reaction, since this a shorter time-constant. So we changed it to the original value for curiosity, and let the set rest.

Schematic Changes

Unfortunately, changes are necessary. Manufacturers are faced with necessities and certain dictated changes once in a while and there's no help for it. Our problem (with all past waveforms still correct) was transistor Q402 on the PW400 deflection board. This transistor monitors the difference voltage, and, therefore, the current across R425, a 2 percent resistor. Under good working conditions TP1 measures some 44 volts and TP2 about 54 volts. Since this is a 3K resistor, and 10 volts is dropped across it, the current is 10 divided by 3 x 10^3 or 3.33 milliamperes. Observe that zener CR409 should be an emitter reference of 10 volts also; consequently, the base-emitter potential is zero and the transistor is **not** conducting.

Should the high voltage current increase, however, the voltage across R425 will also increase, forward biasing Q402 into conduction, but not saturation, and therefore NOT a severe 10-volt clamp. Since C417 removes all AC from the base of Q402, the forward bias would be straight DC. And this means that when Q402 begins to conduct, current flows through the collector and base into the emitter and across R429, causing a subtractive voltage drop in the 32 volts at HH (Fig.14-14) and throwing the horizontal blocking oscillator out of sync. (You might also say that any two positive voltages subtract.) So that's the story of the reluctant oscillator. CR401, by the way, can be involved here also, but undoubtedly in a much more decided way, since we doubt the oscillator would swing back into sync of its own accord. What happened here was that the transistor beta dropped from 200 to 20. Since it is

Fig. 14-15. Sencore's TF17 circuit transistor checker is handy for FETs and bipolars when you're in semiconductor trouble.

a small transistor, and being partially on, it became more of a resistor than a transistor, and pulled the oscillator's DC voltage down so that the set had to run a bit before sync recovery. A transistor checker (Fig. 14-15) is very handy here when you aren't always sure. Also, diode CR409 must be very nearly 10 volts, not 9.2 or 9.5, or you're in real trouble! Later CTC46 chassis include this diode-transistor modification to prevent runaway high voltage and, thereby, remove any possibility of unwanted radiation.

The best way we've found to check the HV regulator action is to monitor the voltage between the emitter and base of Q402. With a 68K R428 (ours measured 75K and we left it in the circuit), this voltage should be 8.5 and 9 volts either at the two electrodes of the transistor or as the drop between TP1 and TP2. Adjust the high voltage carefully with the brightness down, then recheck. If advancing the brightness control causes the Q402 base voltage to rise above its emitter, then the regulator is not operating as it should, and you should try a replacement.

The zener may be bad, too, so monitor both electrodes simultaneously. The regulator transistor, by the way, is something of a power transistor and will not check accurately except under load, so don't expect an ohmmeter to tell you anything, unless there's an absolute open or short. Use a good power transistor checker every time! This equipment, best used with the component out of circuit, will insure proper loading, sufficient current for a substantial test, and give you a quality indication of what both power and small signal FETs and transistors can really do. An ohmmeter is only a very poor, inadequate last resort.

As for the actual horizontal oscillator alignment, you have to put the first peak about 0.5 volt higher than the second peak just before retrace as shown in Fig. 14-5. Afterwards, check to see that the receiver when turned off does NOT lose sync as the raster is extinguished, but lets the entire picture collapse as originally framed under full power. This latter is the acid test, not only for this receiver but for any B&W or color receiver.

While we're on solid-state horizontal and high-voltage problems, RCA Technical Manager, H. C Horton describes what might have been identified heretofore as output Barkhausen oscillations, snivets, squiggles, etc., in the older tube receivers. Here, we'll just call these various straight and bowed lines that sometimes show on weak signals as high-voltage interference. See Fig. 14-16. In another instance, the author found that the green MAD module had become weak and developed some oscillation that also showed as a squiggle on the right side of the screen. A replacement, of course, cleared the trouble and also brought up the green gun considerably.

POINTERS ON ALIGNMENT

As with troubleshooting, you cannot use antiquated equipment for alignment. A lot of very useful sweep alignment gear has now appeared on the market. There's RCA, Lectrotech, Heath, B & K, Sencore—to name a few, plus some imports. So there is competitive gear from which to pick. But you should make some sort of educated selection, depending on anticipated reliability, accuracy, utility, and what you need to spend. Again, as in selecting an oscilloscope, you're going to get what you pay for. A pinched dollar buys a cheap product, and a "cheep" sweep generator just doesn't do the whole job— not if you're a professional. Top service equipment sweep and markers cost between $400 and $500 (round figures), and there's no way around it. Combine this with $500 to $700 for a

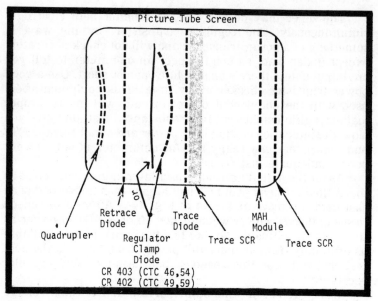

Fig. 14-16. Vertical interference lines are caused by trouble in a solid-state receiver's horizontal sweep system.

really good dual-trace oscilloscope, a crack color-bar generator, and you have between $1,000 and $1,300 invested already, and this is precisely what it's going to take (1972 prices). How do you select a good sweep generator? We can't give you all the answers, but here are some pointers to help.

Voice-coil driven capacitors and inductors are simple, inexpensive, and easily designed. The vibrator can be either a capacitor or coil. Possible problems are caused by too much mechanical vibration, poor linearity, and inductors used as tuners. Needed is a heavy duty voice coil magnet and small diaphragm movement for best results.

Saturable reactors, the "increductor" types, are still used to some extent to produce IF and VHF frequencies. They are both simple and reliable over their limited range. However, saturable reactors often have low Q and are subject to possible drift problems and residual FM developed from surrounding magnetic fields. On the other hand, there are no vibrations because the sweep is electronic, and very slow sweep rates are practical. The output usually is both flat and linear, if the design is good.

Solid-state generators with saturable reactors and varactors have little drift, almost nonexistent heating

364

problems, frequency ranges from low MHz to microwave frequencies, and individual frequency oscillators, or several, depending on the sweep ranges needed or desired. Combined with switch-activated crystals at close to .005 percent accuracy, this type of generator is highly desirable. However, if a generator has individual crystal-controlled RF oscillators, the channel selection will often be limited. And if oscillators are varactor tuned, there may be some spurious marker beats here and there that you must learn to recognize. Naturally, the warm-up time is very short, and the crystals and the oscillating circuits seem to stabilize rapidly. The overall flatness of the sweep output is a good thing to see, as is the range of most generators.

Markers are one of four types (Fig. 14-17)—birdie, absorption, pulse, and intensity. Pulse markers sweep (or ring) a crystal-tuned circuit; birdie markers mix a sample of the sweep signal with the crystal frequency output for a beat; absorption markers cause a frequency suckout on the swept waveform; and intensity markers are applied to the Z axis input on an oscilloscope's cathode ray tube. For accuracy, you must have all crystal-controlled markers. A general slide-rule or "machined" metal dial is not sufficient—it can't be! Most generators, by the way, provide birdie markers, and these are fine if you can control the amplitude and width completely, and if you can turn them 90 degrees (sideways) for the steeper portions of any well-trapped response curve. You'll see why shortly.

The Sencore generator (Fig. 14-18) uses a birdie bypass varactor sweep system, and you'll see both vertical birdies and horizontal birdies.

Do you really need to align? That's the biggest question that has to be answered. What are the criteria?

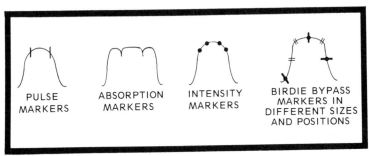

Fig. 14-17. Various types of markers are used. Markers must be added to the response curve only **after** the sweep has passed through all tuned circuits to prevent distortion.

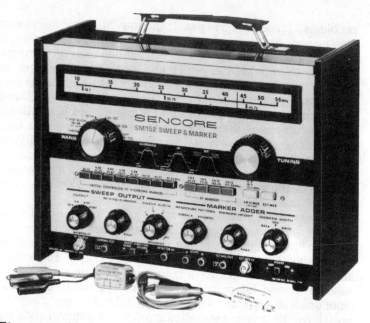

Fig. 14-18. Varactor-tuner SM152 Sencore sweep, crystal-controlled marker generator with ranges for 10 MHz to 900 MHz. It contains 13 close frequency crystal markers with both vertical and horizontal markers, virtually flat output.

Does the receiver have good picture definition, but just weak?

Picture quality is poor, but the fine tuning has little video effect?

Some channels are OK and others poor?

Vertical or horizontal bars in the picture?

Bad horizontal or bad vertical sync?

Good video but sound bars in the picture?

Dark picture, good detail, brightness or contrast unchanging?

Colors washed out, video good?

Color fringing, sound and picture OK?

If the receiver has any or all these faults, **you don't need to do an alignment! But you must look at:**

Video amplifiers, AGC, or RF amplifier.

If the fine tuning won't take the sound bars out, tune the 4.5-MHz and-or and 41.25-MHz sound traps. Or, you may have a component breakdown.

You have horizontal or vertical troubles, usually, if the sync is bad one way and not the other. If sync is bad both ways, of course, it can mean anything from AGC troubles to a faulty sync separator, etc.

A dark picture has to do with AGC, overloaded video amplifiers, insufficient high voltage, or a shorted picture tube.

Poor colors are probably the result of bad chroma amplifiers; mixed up colors can be trouble in the demodulators; no colors may be trouble in the 3.58-MHz oscillator, or a lack of chroma bandpass response.

For color fringing trouble, check the static and dynamic convergence.

A smeary picture without sound interference often comes from the video amplifier and can be due to defective peaking coils, leaky and open capacitors and bad load resistors.

You may definitely need alignment IF:

The picture quality is poor and fine tuning **does** affect definition and possibly contrast.

Poor, grainy picture, touchy sound, wavery sync.

The IF link may not be aligned in older receivers after a new tuner is installed.

New IF amplifiers may have been added or circuits repaired.

Check a test pattern for low- and high-frequency response and leading or trailing edges. As scenes change during a broadcast, distortion will vary.

Weak, watery, unsaturated chroma signals, and a poorly defined picture.

A 3.58-MHz oscillator perpetually falling out of sync could be caused by oscillator or burst amplifier problems. And while we think of it, tune your 3.58-MHz color subcarrier from a broadcast signal, then check another station or two to be sure the lock in range is right. The burst amplifier feedback circuit can also be tuned the same way. Every hue control should be able to go from magenta tones on the left side of the screen to greens on the right—at least 30 degrees in either direction from the mechanical center position (the chroma bars that start with yellow-orange and go to greens on the (right).

Now you should be ready to look at an IF sweep and vectorscope alignment.

VIDEO IF ALIGNMENT

In any alignment, you begin by setting up the sweep-marker generator (Figs. 14-19, 14-20) and a good sensitivity

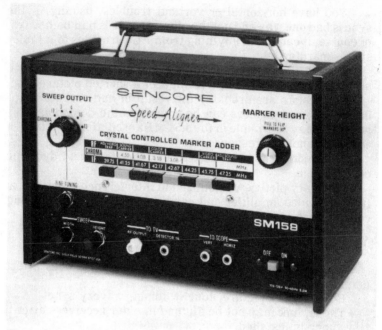

Fig. 14-19. Sencore's SM158 Speed Aligner for VHF TV RF, IF, and chroma alignment.

oscilloscope (Figs. 14-21, 14-22 or 14-23). When alignment as we know it now is outmoded, you may be using a sweep and function generator like the EXACT Model 7060 (Fig. 14-24) or 124 multigenerator for IF and stereo checkouts.

If your equipment has 3-pronged plugs (including ground) and your bench electrical setup has a common, so much the better. If not, securely run braid (coaxial cable with Bx connectors is good) between (or among) all equipment, including the television receiver, and if possible, pick a single point on the TV chassis for all grounds to meet. Then select a bias supply either from the TV itself (through a potentiometer), or an isolated supply such as one contained in the B & K sweep generator or the Sencore BE156 separate bias source. Make all hookups precisely as specified by the manufacturer. Sometimes a bias voltage will vary somewhat from that given, but when a manufacturer asks for an accurate peak-to-peak display on the oscilloscope, he has a definite reason, and you must comply. So again, your scope must be accurate!

In this particular receiver, (Fig. 14-25), we had to remove the tuner-IF cable, and the MAH, PM200, MAL, MAC, MAE,

Fig. 14-20. B & K (Dynascan) Model 415 solid-state sweep-marker generator with RF on Channels 4 and 10 and lighted marker signals. Includes bias supplies, 400-Hz modulation, V and H markers and 14 crystals.

and MAN modules, ground certain points, shunt two others with a capacitor, plus warm up the equipment and receiver for 10 minutes before beginning. First there was the interstage alignment, then the overall IF alignment.

The initial overall response we picked up is shown in Fig. 14-26. The receiver, of course, was working nicely and an alignment was not indicated. On the left baseline is the 47.25-MHz adjacent-channel sound trap marker, on the left slope is the 45.75-MHz video carrier, at the top the 44.25-MHz center point, on the right slope the 42.17-MHz chroma carrier and on the right baseline is the 41.25-MHz sound trap. The chroma carrier is almost exactly at the 50-percent point, but if you want to be a little persnickety, the video carrier is about 60 percent up on the left slope. Do you align or don't you? Under ordinary circumstances with the receiver running well, probably not, but for demonstration purposes we did touch it up. With a twiddle here and there, the response curve came out to be that in Fig. 14-27—not quite as symmetrical, but precise. The markers, right to left, are 39.75, 41.25, 42.17, 44.25, 45.75, 47.25 (all MHz), and all exactly in place. The only difference is a slight out-of-round on the left top of the haystack, and we can always live with this if all the markers are right. For a precise check, we turned the markers (Fig. 14-28) on their sides. Notice that the vertical baseline trap markers now disappear. The peak-to-peak amplitude is a total of 5 volts.

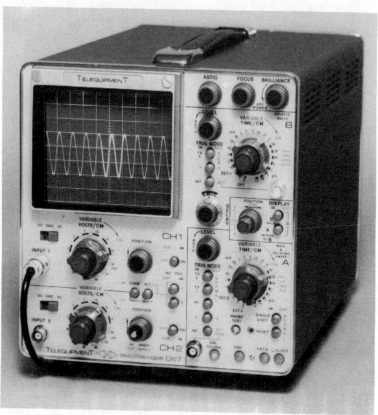

Fig. 14-21. Tektronix deluxe D67 Telequipment scope with delayed sweep and delay line, two time bases, regulated power supplies. Your own scope doesn't **have** to be this fancy, but it helps. Accuracy is better than its guaranteed 3 percent.

CHROMA ALIGNMENT

Again, we were instructed by RCA to remove modules MAH, MAK, PM200, MAL, and MAN, add another short, and also a resistive coupling-termination (no AGC). This time we set the sweep generator to the pre-set **chroma** position so there would be no tinkering with the sweep, except the height and width of the markers. The initial chroma response found is shown in Fig. 14-29. Again, this is excellent and certainly not worth disturbing. The author wanted to investigate, however, and so we began a touchup. Also, and simultaneously, we used a vectorscope with gated rainbow generator to confirm (or

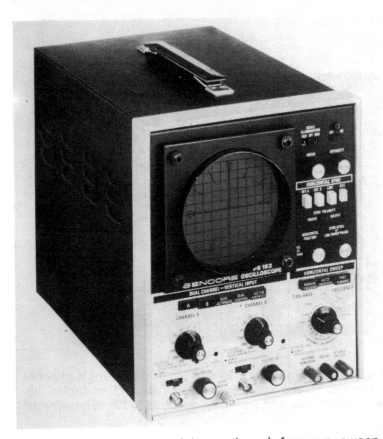

Fig. 14-22. Dual-trace, triggered and free-run sweep, Sencore PS163 scope with front panel vector inputs; deflection factor to 5 millivolts, sweep to 100 nanoseconds.

deny) our findings. This initial display is shown in Fig. 14-30, with the brightness advanced to normal. In the reduced brightness photo in Fig. 14-31, the waveform is relatively clean but distorted, especially in the first three petals, and there are some phase problems in others. As you can see in Fig. 14-32, we connected an 80-mfd microfarad capacitor to AB on PW600 to take out the luminance channel, and the vector pattern looks better.

To continue with the alignment (Fig. 14-33), the 4.08-MHz marker is on the left slope, the 3.58-MHz marker in the center; the 3.08-MHz lower sideband marker is turned off. This looks pretty good, and is within the left downward slope; in this instance, toward a **lower** frequency. We do the same thing for

Fig. 14-23. Dynascan's B & K Type 1460 single-channel triggered sweep scope with 10 mv deflection factor, 10-MHz bandpass, and sweep 0.5 seconds to 500 nanoseconds.

the 3.08-MHz marker, but show what happens when you turn one of the transformer slugs a bit too far. The marker begins to go down the slope of the curve, narrowing the bandpass—and that's not good (Fig. 14-34). So we reduced the amplitude and width of both markers and put them at the top, just inside the specified slope limits (Fig. 14-35). Next, we increased the amplitude and width of the markers so you could compare them with the initial picture (Fig. 14-36). See any real difference? I doubt it, but look at the final vectorscope picture (Fig. 14-37). Really, all we did was to reduce the chroma amplitude and turn the hue control until the third bar and the ninth bar were pointed to +R-Y and -(R-Y), 270 degrees, respectively. The pattern now tells us this receiver is demodulating at an angle of 105 degrees, the first bar starts at 40 degrees, instead of the "perfect" 30 degrees, and there are no crossovers in the display to indicate transformer mistuning.

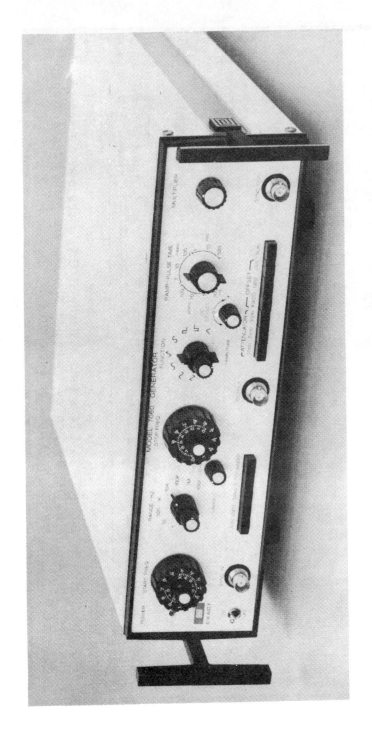

Fig. 14-24. EXACT Model 7060 VCF-Sweep generator with ranges from 0.0001 Hz to 11 MHz.

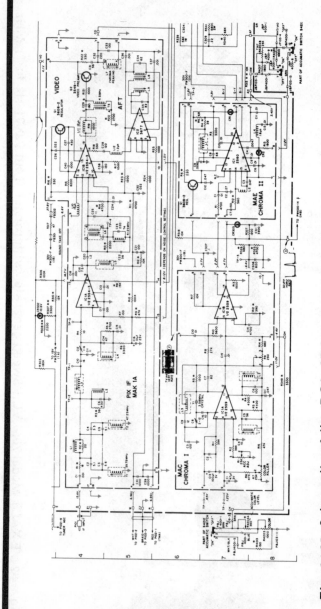

Fig. 14-25. Schematic of the RCA CTC46 (and basically CTC49) video and chroma plug-in modules. The amplifiers represented by triangular symbols are really integrated circuits. (See TAB RCA Color Manual, No. 578, for an explanation.)

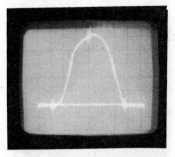

Fig. 14-26. CTC46 initial overall video IF response.

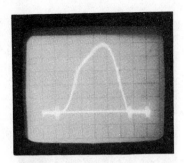

Fig. 14-27. Final IF alignment curve. Markers left to right in MHz are: 39.75, 41.25, 42.17, 44.25, 45.75, 47.25.

TUNER ALIGNMENT

In any tuner alignment, you're likely to be told to add so much AGC DC to the tuner input, additional AFT voltage, and use certain channels for best results. You then must tune certain trimmer capacitors, squeeze a few coils, and be sure the local oscillator is on frequency so that the center IF output is 44 MHz, plus even a bit of trap twiddling. The instructions for each tuner will vary somewhat, but this is about what you will see. Occasionally, you're given two settings, say, Channels 6 and 13, and the others are supposed to fall in place automatically. Then you may be told to add specific terminations, and connect an oscilloscope to certain spots on the tuner to observe the outputs.

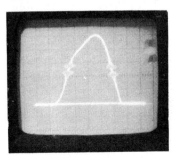

Fig. 14-28. Same as Fig. 14-27, but with the markers turned on the side for a precise check.

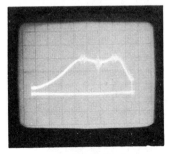

Fig. 14-29. Initial chroma waveform with 3.08 MHz on the right, 3.58-MHz in the center, and 4.08 MHz on the left.

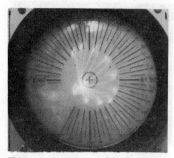

Fig. 14-30. A normal brightness setting on cathode-fed CRTs plays havoc with a vector display.

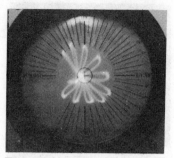

Fig. 14-31. Vector display with the brightness turned down.

It is important that you do what the instructions say to the letter, for there's no magic way of executing an alignment without following the manufacturer's instructions; most of us who have tried otherwise have found out rather decisively the error of our ways. However, there are quick checks which you should know about.

In the Zenith tuner 19CC19 (formerly 4B25C19), we simply connected the Sencore SM152 sweep generator to the antenna terminals and applied a 3-volt bias to the tuner AGC input, with the receiver AFT leads disconnected. The scope was connected to the tuner IF output through a demodulator probe. First, we found a channel where we could produce a reasonable pattern (Fig. 14-38). This Channel was 4, and on the right top of the curve—where it should be—is the 67.25-MHz video carrier, and on the left is the audio carrier represented

Fig. 14-32. Vector display with an 80-mfd capacitor connected to AB on the PW600 mother board.

Fig. 14-33. The 4.08-MHz marker is close to the left edge; 3.58 MHz is still in the center.

Fig. 14-34. The 3.08-MHz (right) marker shows misadjustment. It is not inside the top of the curve.

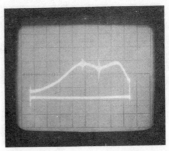

Fig. 14-35. Small markers help make precise adjustments in close chroma alignment.

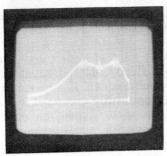

Fig. 14-36. Once critical adjustments are made, the amplitude of the markers can be increased.

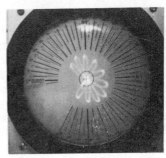

Fig. 14-37. Final vectorscope pattern **after** sweep alignment, an excellent verification of both procedures.

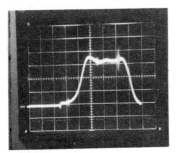

Fig. 14-38. Channel 4 response curve showing both markers. Video is on the right.

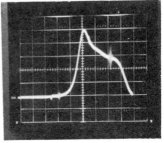

Fig. 14-39. Incorrect bias distorts the response curve.

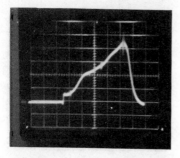

Fig. 14-40. Curve showing the effect of improper fine tuning.

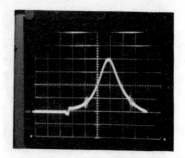

Fig. 14-41. Poor Channel 10 response, but markers are firmly in position, so try another more favorable channel.

by the beat of a separate 4.5-MHz crystal. This, of course, is a very pretty tuner double-humped pattern exactly what you should see on every channel.

Suppose, however, you weren't sure of your bias. The pattern might look like Fig. 14-39—a sloppy, slanted waveform that is completely useless. This condition can also happen if each piece of your equipment isn't uniformly grounded to the receiver and to each other. On the other hand, you could develop somewhat the same condition with misadjustment of the local oscillator fine tuning (Fig. 14-40). If the output of the sweep generator were too high, the curve would "bump" a certain upper level and then begin to flatten the pattern out. So you must always allow a little leeway for bias and generator amplitude adjustment so that the waveform isn't distorted. (The manufacturer usually specifies the amplitude.)

Let's assume you now have a complete check on Channel 4, and now go to Channel 10 (Fig. 14-41). Look what happens here: you almost have an IF haystack, but with the 193.25-MHz (right) and the 4.5-MHz markers much too far down on the response curve. What do you do? Just go on to a higher channel and hope for the best. The final channel turns out to be 13, and this curve doesn't look too terrible (Fig. 14-42), but it does have an extra marker on the top that shouldn't be there. When you have such a situation, simply reduce the amplitude of all markers and-or turn off the controlled markers one-by-one until the spurious marker stands alone. In this instance, you will find the spurious marker (Fig. 14-43) on the extreme left when the 4.5-MHz marker is turned off.

All these are pretty good lessons in alignment, because, if nothing more, they tell you what to prepare for. Following

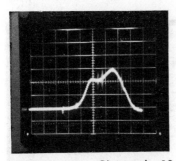

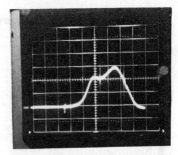

Fig. 14-42. Channel 13 response curve—better, but with a spurious marker on the left.

Fig. 14-43. With the 4.5-MHz marker cut off, the spurious one is on the left and is immediately recognizable.

exact instructions is no more than a matter of reading a rather simple book. The quick checks described will help you to discover the good or bad of a tuner, IF, or chroma section before, we hope, you dive into a full-fledged alignment, and a quick check can go a long way towards telling you what may or may not be wrong. Use it whenever, in doubt, especially if you have a good sweep generator and handy bias supply. Sweeping a tuned circuit can tell many things, both in alignment and when searching for troubles. For instance, if a certain IF or RF stage can't be tuned, there is something obviously wrong with that tuned circuit, and the sooner you correct it the quicker your difficulties will be over. Another evergreen suggestion is: **set the traps first**, then go about any needed alignment.

IF ALIGNMENT PROCEDURE

Ordinarily, you set the Channel selector to 13, connect the sweep generator input to the base of the third IF, the scope to the emitter of the first video amplifier, and begin adjusting the fourth video IF transformer. Afterwards, connect the generator to the tuner test point, and do the rest of the trap and transformer settings. But, again, there's usually a way to quick check the IFs. Apply external AGC to the first IF, and if your generator pads are near the proper impedance, you can simply connect the RF input of your sweep generator to the tuner-IF coupling network. Naturally, the scope connects to the video detector (if you have ancient equipment) or first video amplifier. In the Zenith 19CC19 (Fig. 14-44), the RF input could go between L101B and L101A, and the response curve

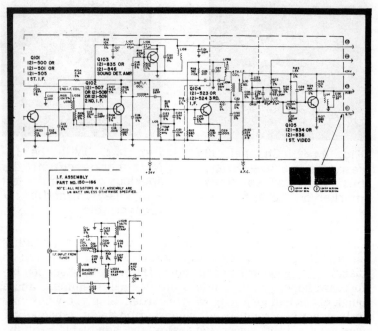

Fig. 14-44. Schematic of Zenith's 4B25C19 (now 19CC19) plug-in video IF strip.

would be picked up across R127 in the emitter of first video amplifier Q105. The resulting waveform (Fig. 14-45) should be a true representation of the video IFs. We extended the amplitude of the markers so you would have no trouble identifying them. From left to right they are: 47.25-MHz lower adjacent sound carrier trap video carrier, 45.75-MHz, center frequency of 44.25 MHz at the top, 42.17-MHz chroma subcarrier crystal beat on the right slope, and the 41.25-MHz sound carrier at the right baseline. This is an excellent IF response. The curve could be moved a hair to the right, but it might upset the video and chroma carrier positions; therefore, there's no need to change it all.

We're told that Zenith aligns tuner and IFs separately at the plant, and only touches up the converter coil when checking the final mating. Not too long in the future, there'll probably be no actual touchups at all.

CHROMA ALIGNMENT PROCEDURE

Happy to say we couldn't show any simple alignment of the chroma circuits in this Zenith with a sweep generator if we

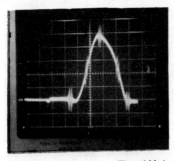

Fig. 14-45. Zenith's 19CC19 video IF response curve. Excellent marker positions; no need to change them.

Fig. 14-46. Vector pattern indicating a need to adjust the receiver's contrast or brightness.

wanted to. There are no tunable transformers, and the only variable passive components are resistors. So, with these receivers, which are self-bandpass limiting, you simply use a voltmeter (or DC scope) and a color-bar generator, and the chroma alignment is done—just about as quickly as we can write it.

But we did think a few pointers on chroma alignment with a vectorscope, even using this receiver, might be of benefit. In one of these receivers, where chroma and video are mixed before entering the picture tube, you usually have to adjust brightness and sometimes contrast, or you'll have a picture like Fig. 14-46. Later, lower the brightness, but fail to AC short the luminance channel and provide no ground for the vectorscope, the unpleasant "barrel" such as shown in Fig. 14-47 will appear. Then, if you're using one type of color-bar generator, the pattern you'll see appears as in Fig. 14-48, and this is a well-known generator, too. But it's so smeary you can't use it. In Fig. 14-49, if the fine tuning is slightly off, or should the bandpass transformer be somewhat misadjusted, the petals in the pattern tend to become rather wide open. On the other hand, a badly tuned crosstalk potentiometer (and this "pot" controls both passband and amplitude of the chroma input) can cause highly undesirable crossovers that positively cannot be permitted (Fig. 14-50).

What should a good vectorscope pattern look like? Fig. 14-51 shows good chroma alignment using our own modified Mercury 1900 color-bar generator that is virtually free of 189-kHz oscillator "gating" spray in the chroma oscillator. The first bar begins at about 30 degrees; the second is obviously 60 degrees; we've set the R-Y third bar at 90 degrees, resulting in

Fig. 14-47. This pattern appeared with no ground and no capacitor short.

Fig. 14-48. One available color-bar generator produced a pattern with considerable smear.

a demodulation angle of 105 degrees between the third R-Y and sixth B-Y bars, with all others falling nicely into place as they should. You'll recall that the color generator gates every 30 degrees from burst reference around to 300 degrees. So this chroma circuit, as you can plainly see, is highly accurate, and offers excellent color demodulation. Vectorscope connections were made at the limiting resistors going into the base of the cathode ray tube. You might also remember that this is a cathode-fed tube, and vectorscope inputs must be 180 degrees out of phase with those in a grid-screen CRT matrix receiver. An R-Y (RC) equalizing network was used to help make the pattern somewhat rounder and to pick up the high frequencies. Values were 390 ohms shunted by 180 pf. The vectorscope used is a Mercury 3000.

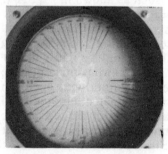

Fig. 14-49. The crosstalk adjust spreads the petals and changes the input bandpass.

Fig. 14-50. Crosstalk can also cause bad crossovers, and this means washed-out colors.

Fig. 14-51. Pattern achieved when the Zenith 19CC19 chroma circuits are aligned correctly.

AGC PROBLEMS

To acquaint you with AGC problems, let's run through some of the day-to-day difficulties briefly. Since most receivers today have a keyed AGC system, anytime you have high-voltage problems, the keying part of the automatic gain control function is not there, and sync can feed through to the audio, causing a buzz. Whenever there is AGC overload, you may or may not hear this same sync buzz, but the screen will certainly become dark, and the picture may go into saturated bends. These bends occur because the tips of the sync pulses have been clipped by amplifier overload, and too much overload will cause the image to lose sync entirely, beginning with the vertical oscillator.

In tube-type sets, AGC is a negative voltage that cuts down on grid conduction. In most transistor receivers, all AGC is a positive forward bias that causes a reduction in hfe (forward current) in specially selected transistors **before** these transistors saturate and lose both gain and power in the process. With AGC bias, each controlled stage continues as an amplifier, but stays out of saturation unless heavily overbiased.

Of course, there is little or no AGC if the keyer fails, or if there is no video input to the AGC to establish its conduction level. The usual collector or plate DC voltage probably will provide a little action, but not enough to gain a satisfactory picture. A lot of alignments, by the way, have been done for just such a flimsy reason. Where there is the slightest doubt, clamp the AGC line in vacuum tubes with a negative source and no disconnects, and in transistorized IFs, break the AGC line somehow and apply a positive voltage from a stiff DC source and see if there is any worthwhile response. If not, your problems are probably in the IFs, and then you need not only an AGC source, but a demodulator probe for IF investigation as well. These serviceman's demodualator probes, incidentally, rarely give you true peak-to-peak values, but are used solely to monitor signal passage and relative stage gain among the two to four video amplifiers.

HIGH-VOLTAGE vs LUMINANCE AMPLIFIERS

Since high voltage (keying) and luminance amplifiers (video) play fundamental roles in operating the AGC, it's natural we should consider these two functions as well. Many servicemen are perpetually confused when they look at a color receiver and see a blank screen. The immediate inclination is to tear into the set, replace anything in sight, and hope for the best. Obviously, this is absolutely not the way to approach this problem.

First, measure the high voltage to discover if there is any. If you have high voltage, then proceed to the video detector and follow through to the cause of the problem. Frankly, it's just as simple as that. Or you can use an oscilloscope all the way. Look at the conduction period of the high voltage (if there is any) without ever going inside the set; you'll have to be near the HV cage or transformer though, so determine if it looks good. Should this be your decision, **then** go to the limiter and video channels and find out what's wrong.

We deliberately didn't mention the chroma-video outputs on the new CRT cathode-fed receivers, because usually only one of these fail at a time, and the colors **and** luminance are then promptly affected, according to the extent of the trouble.

VERTICAL JITTER; VERTICAL LINES

Another mystifying situation is vertical jitter or a vertical line problem that affects some receivers. Unfortunately, the cause of the trouble can range from a bad tube, transistor, dirty potentiometer, or possibly a slightly leaky capacitor (but not likely). It is a very difficult condition to diagnose in that horizontal frequencies are filtering through the deflection yoke and modulating the vertical output. Remember, vertical line problems can come from horizontal modulation, and horizontal problems from vertical modulation. Sounds like the pincushion transformer, doesn't it? Yet, we've seen times when a dark line on a CRT would appear with diminished brightness and, in normal control settings, the vertical scan would just sit there and ever so slightly jitter. When we went to the vertical stages, numerous unknown frequencies appeared on the drive to the vertical output. And the condition was even worse at the plate of the vertical output tube itself. With a triggered sweep, we quickly determined that these stray frequencies were horizontal pulses leaking through the deflection yoke. A yoke change put conditions right immediately.

Another condition we recently ran into was a "waterfall" problem, where the vertical sweep was modulated with what looked like low-frequency wiggles. You'd be safe in betting either a vertical filter or the vertical deflection yoke had a lot to do with this. The receiver—a new Motorola—had no other source of 60-Hz since its starting voltage is a full wave doubler and its horizontal oscillator-run power supply is, of course, cycling at 15,734 Hz. Sometimes, in tube sets, you might suspect a damper, but this receiver is damped on a plug board, and that one had already been substituted. A crosshatch generator oftentimes can help diagnose an aggravated sync fault. Just mentally think horizontal for vertical, etc., and the extent of the problem should take shape. The solution, of course, could take a little longer.

All sorts of conditions can occur in any receiver. Many will appear to be similar to many you've seen before and yet turn out to be something entirely different. Background experience is excellent, and a few corrective tips from the manufacturer often are a windfall. But in most instances, you have to depend on your own initiative. When you have a color problem, look at the chroma and-or luminance outputs (if they're together, or separately, if not) with an oscilloscope and color-bar generator and trace back to your troubles. If video-chroma voltages are good and the waveforms true, by all means use a cathode ray tube checker to prove whether or not the CRT is shot. There are some good CRT checkers around: two we know are Sencore and Mercury (Figs. 14-52, 14-53). The Sencore is slightly more advanced, but both will **track** each of the three RGB guns relative to each other, and this is the true test of whether or not a color tube will actually perform, provided it has initial emission and no shorts or opens. CRT testers, good ones, are valuable investments and are enormous time-, money-, and reputation-savers.

COLOR FAULTS

To illustrate the visible effects caused by trouble in chroma circuits, we've used an almost totally integrated circuit TV color section. No longer do you have the simple barber-pole faults of oscillators being off frequency, single colors absent while all others shine, a potentiometer or quarter-watt resistor being burned or changing in value from hot or warm running currents, or big capacitors opening or shorting and producing large faults that are fairly easily dealt with.

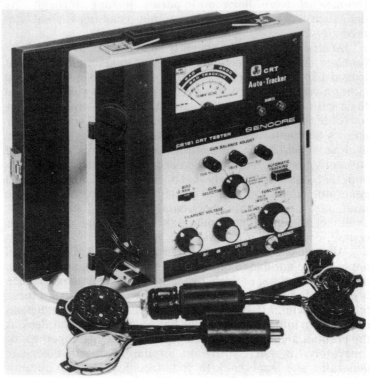

Fig. 14-52. Sencore's CR161 CRT checker with automatic gun tracker (memory) for all tubes, including Trinitrons.

Fig. 14-53. Mercury's 880 is easily set up and is a color tracker as well as a rejuvenator.

In IC color processors (and before long, all-IC receivers), a host of small, even minute, breakdowns will completely blot out color altogether, pop out the circuitbreaker and turn off the entire set, or shift the phase colors so that the problem appears to be one caused by anyone of a number of sources. With ICs in chroma circuits, you have DC-controlled bandpass amplifiers, DC-phase shift hue and chroma controls, filters of all descriptions, and tuned circuits that must maintain their parallel and series resonant frequencies as maximum and minimum impedance paths, respectively. The best of professionals are designing these receivers with programmed computers to help, and semiconductors are not nearly as restrictive as vacuum tubes, considering what can be done with them. So with ICs, you have but two salvations: a handy socket mount and-or plug board, and a very good quality DC triggered sweep oscilloscope. Thereafter, it will be a case of signal in, signal out, and checking rated voltages from the set's power supply. Figs. 14-55 through 14-78 in the color photo section depict the screen effect of various chroma circuit troubles. (See pages 33 through 36.)

Regarding color-bar generators, positive dot and crosshatch pattern stability without the use of J-K flip-flops

Fig. 14-54. Bell & Howell FTb Canon camera used to take the photos in the color section. It has an FD 50mm F:1.4 lens. (Courtesy Mike Laurane, Bell & Howell, Chicago)

387

and other ICs in the count down chain will be either extremely difficult or downright impossible to maintain. The IC generators may cost a little more, but the price stretched over the months will be wholly worth the needed professional convenience. Once adjusted, these count-chains stay and produce fine-line, highly stable dot-bar patterns that make convergence on the better receivers a pleasure. You will also be able to set up the poorer converging sets so that fringing barely shows along the sides, top and bottom, if at all, from a distance of six or more feet. In the pictures you'll see several examples. The only problem you continue to face with available color-bar generators is they still don't gate the chroma oscillator cleanly enough to avoid producing fuzzy vector patterns. We've made the point now to enough test equipment manufacturers so that some day one will see the light and proudly offer something with more utility than we have now.

QUESTIONS

1. How can you justify $500, $700, $1,000, $2,500, for a good oscilloscope?

2. Is there ever a time when an analog meter can be used in place of an oscilloscope?

3. What do you think will soon replace vane-type analog meters?

4. If you're using a scope's AC amplifiers and want to know the DC voltage the AC wave is "riding" on or "hanging" from, what do you do?

5. Can you calibrate a DC oscilloscope with an AC voltage?

6. What are convenient time-base rates to observe vertical and horizontal sync waveforms?

7. In the waveform in Fig. 14-8, what's all the jitter in the top pulse? Is this a satisfactory condition?

8. Why can't you use the dual-trace chopped mode on traces at kHz frequencies and above?

9. Why don't you randomly change parts just to see if the set won't "work"?

10. Are solid-state color receivers susceptible to Barkhausen oscillations?

11. From reading the descriptions, what sweep generator would you select if you wanted 1) reliability and restricted sweep coverage; or 2) a wide range and sweep flateners?

12. Name the four types of markers.

13. If a receiver has good video but sound bars in the picture, does it need video IF alignment?

14. What if you have a smeary picture with no sound interference? Does this indicate a need for alignment?

15. What about IF alignment when the picture quality is poor and fine tuning **does** affect definition and probably contrast?

16. How can you pull the sync and sweep circuits out of a receiver and still get an alignment pattern.

17. Look at Fig. 14-28 and see if side-appearing markers aren't thoroughly useful. But why two sets?

18. Why do you remove luminance when you want to use a vectorscope on a totally cathode-excited picture tube?

19. A chroma alignment (Fig. 14-33) can be best done with all three markers visible simultaneously?

20. What can cause tilt in RF and IF swept waveforms besides bad oscillator tuning?

21. What would you do if, by touching your hookup cables, the shape of the alignment response curve changed?

22. When there are spurious markers what do you do?

23. Do most demodulator probes render accurate p-p voltages? What is their big response problem?

24. How would you determine whether or not horizontal frequency signals are interfering in the vertical circuits, or vice versa?

25. Why do you need a cathode ray tube checker when the outside technician "eyeballs" a gassy appearing tube and declares that it's shot?

Chapter 15

Domestic CATV Systems

With cable television serving 5.5 million homes already in this country through 2700 systems and, by 1975, 1.7 million more are expected, this industry's size and potential is already formidable. In the not too distant future, two-way communications to and from many CATV subscribers' homes may well be in operation, offering shopping services, computer links, banking, schooling, trend polling, and even employment to those who wish such services.

For a good look at CATV and what engineers and technicians are going to contend with, we went to Philadelphia, Pa., and talked with Jerrold Director of Technical Operations, Mr. Caywood C. Cooley, Jr. First we asked, "How do you go about setting up a CATV facility?" And this, generally, was his reply (Figs. 15-1 through 15-4).

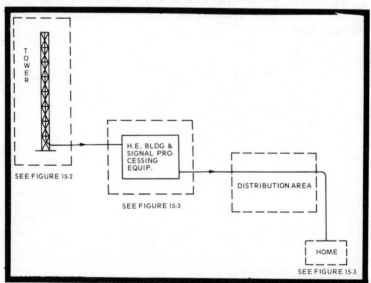

Fig. 15-1. "Standardized" type CATV system showing major blocks with details spelled out in Figs. 15-2, 15-3 and 15-4.

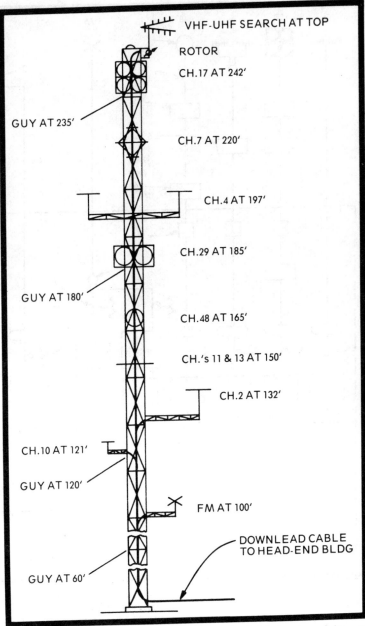

VHF-UHF SEARCH AT TOP
ROTOR
CH. 17 AT 242'

CH. 7 AT 220'

CH. 4 AT 197'

CH. 29 AT 185'

GUY AT 235'

GUY AT 180'

CH. 48 AT 165'

CH.'s 11 & 13 AT 150'

CH. 2 AT 132'

CH. 10 AT 121'

GUY AT 120'

FM AT 100'

DOWNLEAD CABLE
TO HEAD-END BLDG

GUY AT 60'

Fig. 15-2. Drawing of a typical 250-foot high guyed tower designed to support receiving antennas for various channels.

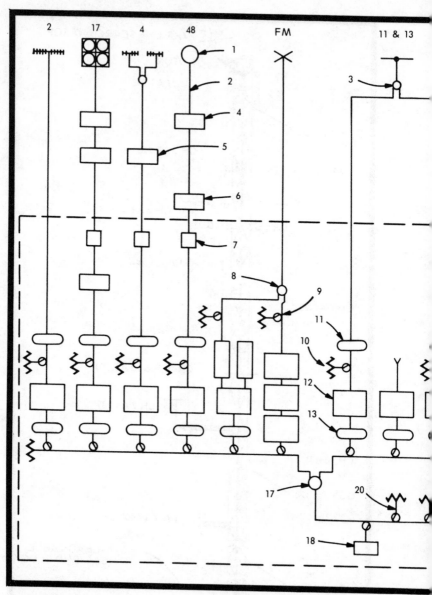

HEAD-END EQUIPMENT LEGEND

1. ANTENNA
2. DOWNLEAD CABLE
3. SPLITTER-COMBINER
4. PASSBAND FILTER (UHF)
5. PREAMP (VHF OR UHF)
6. CONVERTER (UHF TO VHF)
7. POWER SUPPLY
8. SPLITTER (2 WAY)
9. DIRECTIONAL COUPLER
10. TERMINATOR
11. PASSBAND FILTER (INPUT)
12. COMMANDER

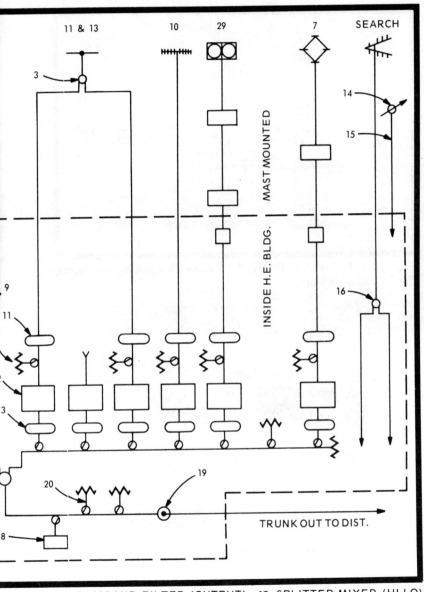

13. PASSBAND FILTER (OUTPUT)   17. SPLITTER-MIXER (HI-LO)
14. ROTOR   18. CARRIER GENERATOR
15. ROTOR CABLE   19. PROBE (TEST POINT)
16. SPLITTER-MIXER (VHF-UHF)   20. TEST POINT

Fig. 15-3. Block diagram of the signal processing equipment in the head end.

393

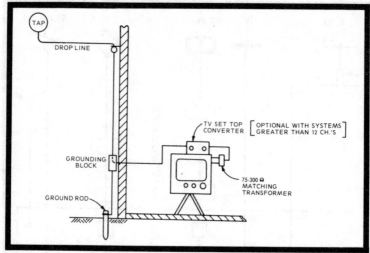

Fig. 15-4. Drop line from the cable tap feeds the home CATV installation.

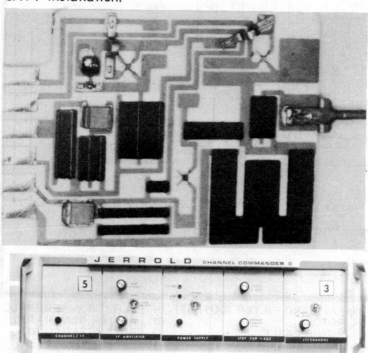

Fig. 15-5. Jerrold Channel Commander 11 signal processor. This one translates Channel 5 to Channel 3.

First you select an antenna site, called a head end, make all sorts of signal measurements from ground level or from overflights in light aircraft or helicopters, then negotiate for real estate. Meanwhile, avoid a site overlooking a city because of neon signs, etc.; pick a spot free from ghosting and differentials between video and sound caused by local propagation conditions.

When these conditions have been satisfied, the receiving antennas are selected. Log periodics are preferred, and are a must on the low-band channels. But on the high band, horizontally stacked yagis can be used to advantage to change pattern lobes, directional pickup, and the like. Next come the preamplifiers, then UHF to VHF converters, with UHF preamps fastened directly to the tower. Downleads are all through wideband static couplers, with cable in aluminum sheathing and either a polystyrene or polyfoam dielectric.

In the antenna shack, the received signals are fed into heterodyne signal processors such as that shown in Fig. 15-5. Video and sound are converted to two intermediate frequencies (IFs), then both are AGC controlled to establish a level and maintain a constant ratio between sound and picture carriers. In a normal broadcast transmission, the sound carrier is about 6 db below video, but in CATV, Mr. Cooley says, "we go 15 db below video," allowing good adjacent-channel rejection in all but the very poorest receivers.

From the IF frequencies, the signal is beat back up (heterodyned) to the channel desired on the cable. Some systems use a single-channel, others a 26-channel tuner, which separates the various modulations on the carrier and puts a number of channels into a single channel through the tuner of a television receiver. If receiver top converters are used, amplifiers must suppress any second-order harmonics by output cancellation to avoid, say, a Channel 2 and a Channel 6 beat, that will fall at 140 MHz and give you problems if you're using that particular frequency. There's also cross modulation (one channel modulating another) which causes what is commonly known as a windshield wiper effect on the desired channel. It is caused by the horizontal blanking bar of the interfering channel when an amplifier is delivering more output than it should.

CATV signals are then transmitted down main trunklines (Fig. 15-6) through feeders, splitters, and finally into line drops and subscriber homes. There are mainline repeater amplifiers positioned every 20 db down from the original signal level, usually about 2,000 feet apart. Repeater amplifier spacing is normally a function of the cable, the splitters, and

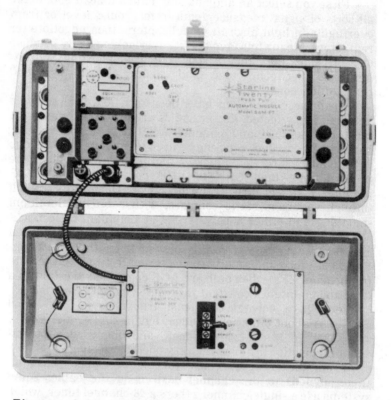

Fig. 15-6. A Jerrold Starline Twenty push-pull fully-loaded repeater amplifier station SP-1.

directional couplers, and the various taps. There are also line bridging amplifiers which supply feeder lines and extenders (Fig. 15-7), couplers, line taps (splitters, Fig. 15-8), and finally the separate house drops.

Do you often find illegal taps, we asked? Sure, replied Mr. Cooley. But on holidays and during summer vacations, schoolboys and girls are sent out to count the number of taps in specific neighborhoods. "Illegals" are than confronted with the evidence and almost all of them take the service thereafter—or are hauled into court.

Are there problems? Obviously. About half the CATV service calls are traceable to the subscriber's receiver. The CATV technician carries a black-and-white receiver along with him and makes separate signal checks besides. He monitors the field strength of channels 2, 6, 7, and 13 at the tap.

If there is a complaint, these checks should certainly uncover the problem if it's a simple one. More complex difficulties usually require supervisory personnel with more elaborate equipment. For instance, main trunk measurements, made at the end of the CATV transmission line, include:

1. Signal-to-noise ratios
2. AGC pilot carrier modulation
3. All signal levels, including a look at the pictures
4. Hum modulation
5. Possibly a sweep response of the entire system periodically every two scanning lines, and in cases of bad problems, a system equalization every 30 seconds. Here, some oversensitive receivers may develop vertical roll during the sweeping period.

Of course, there is also head end maintenance, including tower inspections, cables, amplifier measurements, and possibly, spectrum analysis. With a solid-state system, only one man is needed for each hundred miles of line, whereas with tube equipment a man has to be assigned to each 40 miles of cable.

The outside television serviceman can look for the following problems:

1. Second- and third-order distortion, usually fine diagonal lines in the background that are sum and difference beats. If you want to know the fundamental interfering frequency, loosely couple a signal generator to the receiver's antenna terminals and match the interfering beat.

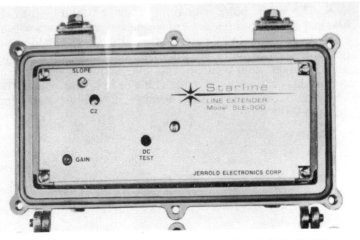

Fig. 15-7. Jerrold line extender, Model SLE-300.

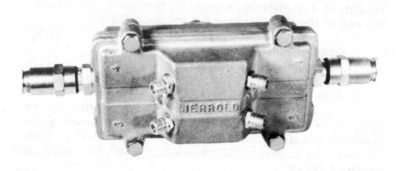

Fig. 15-8. A line tap that will serve four separate sub-scribers.

    2. Modulation envelope delay distortion usually occurs in the head and may appear as leading or trailing ghosts in B&W or color pictures. It may also show as a lack of crispness in the picture. The trouble can be verified by using a good off-the-air signal or, perhaps, a good color-bar generator. If the picture is clean by these methods, the cable is bad.

    3. Where there are microwaves around, whites in the picture can be clipped, causing noisy sound, plus sync clipping that results in vertical rolling.

    4. The color burst signal or color amplitude may be enlarged or reduced because of a heterodyne processor fault or problems in the CATV antennas. Again, use an outside color signal for verification. And, may we suggest, carry a reliable color receiver with you so you can show any subscriber whether his problems are in the cable or in his set. Otherwise, there could be a surprising number of callbacks as your "reward."

Chapter 16
Foreign TV Systems

At last count, there are about eight television systems used in the world, and some have certain features in common: In those countries where electric current alternates at 50 Hz, the field frequencies are 50 Hz and the frame rate is 25 per second. Where the current rate is 60 Hz, fields are 60 Hz and the frame repetition is 30. The maximum video bandpass has been standardized at 3 MHz for the British 405-line system, 4 MHz for the U.S. 525-line system, 5.5 to 6.5 MHz for the 625-line European and French systems, and 10.4 MHz for the 819-line Belgian and French systems.

Frequency allocations for these various systems in the three principal regions of the world are given by geographical location in Table 16-1.

Region 1, The Continent	Frequency In MHz
	47 - 68
Europe, USSR,	174 - 216
Africa, Middle	216 - 223
East, Saudi	470 - 942
Arabia, Iceland	942 - 960
Region 2, The Americas	
U.S., Canada,	54 - 68
Hawaii, South	68 - 73
America, Central	75.4 - 88
America,	174 - 216
Greenland	470 - 890
Region 3, The Orient	44 - 50
	54 - 68
	174 - 216
India, S.E. Asia,	470 - 585
China, Mongolia,	610 - 890
Australia, New	890 - 942
Zealand	942 - 960

Table 16-1. TV frequency allocations in the world regions.

There are also two important and very recent international Geneva agreements that should also be included:

1) In the 470 to 890-MHz UHF band, the frequencies of 620 to 790 MHz have been assigned to television stations in the FM band to receive satellite transmissions. However, these frequencies will **not** now be used in the U.S. by local agreement; only by foreign countries.

2) The UHF frequencies of 2500 through 2690-MHz have been assigned as Broadcast Satellite bands, along with 11.7 to 12.2 GHz.

The standard aspect ratio throughout the world (the ratio of picture width to height) is 4:3, and all systems have 2:1 interlace scanning to avoid the unpleasant effect of picture flicker. FM is universally used for sound modulation, except in older English transmissions and in France and Belgium where AM is the method used. Both U.S. and European systems have standardized on negative picture modulation, where the transmitter power decreases as televised scenes brighten. England, France and parts of Belgium use positive modulation where power and brightness increase together.

Despite the continuing differences in the several European systems, there are now 22 countries in Western Europe and North Africa that are linked by the Eurovision Network, and there are additional links to the Intervision Network of Eastern Europe. Since the launching of Telstar 1 in July 1962, and the later Intelsat series of satellites, distant television transmissions have increased to the point now where, in the Intelsat 4 satellite series, there are channels devoted completely to television. Europe and the United Kingdom have also standardized in the past 20 years, so that in countries using 50-Hz fields, the BBC and ORTF now exchange international programs on the 625-line standard. In Europe, however, there are still the German and French systems where considerable conversion is always necessary, although it is easier to work between these two than among the previous 405, 625, and 819-line systems, respectively, of all three. The remaining outstanding problem between the German and French systems is that of chroma subcarrier frequency tolerance. The German PAL system specifications are normally tighter than those in the French SECAM system, so conversions from German to French are relatively easy, but sometimes rather difficult from French to German.

The PAL and SECAM systems are evolutions of the U.S. National Television System Committee (NTSC) standards adopted in America in 1953. By 1960, the EBU-UER European television group finally decided upon the 625 line 50-Hz system, with a color subcarrier frequency of 4.43 MHz, based on 284.5 times the line frequency of 15,625 Hz. French, German, and U.S. systems transmit luminance and chroma separately, with the usual pair of color-difference signals that are modulated on color subcarriers. The color sync signal is also broadcast on the back porch of the horizontal sync pulse. The following description of the PAL and SECAM systems includes an evaluation supplied by NBC's Jaro Lichtenberg, a native of Czechoslovakia.

The German PAL System

The Phase Alternation Line television method was developed by Walter Bruch in 1961. Here, the phase of the chroma is alternated 180 degrees from one line to the next so it appears out of phase on every other line and therefore cancels differential phase errors. This requires that a line-switching signal must be transmitted also, as well as a synchronizing color burst. The result, though, is that there are few, if any, transmission phase errors, and a hue control as we know it in the United States is unnecessary. Compared to NTSC, the burst does not have the same type phase, but alternates first positive then negative 45 degrees (from 0 or burst) as it is transmitted from line to line, with changes being made by the line switching signal. At the receiver, this alternating burst is decoded and used to sync the burst so that the local subcarrier oscillator is firmly synchronized with the transmitter.

In color transmission, PAL transmits only R-Y and B-Y (since G-Y is a combination of R-Y, B-Y and luminance)— NOT I and Q as in the NTSC system. Both R-Y and B-Y double sidebands with bandwidths of 1.7 MHz below and 1.3 MHz above the subcarrier are transmitted to relieve as much chroma crosstalk as possible.

While NTSC completes its vertical scan in two fields and at 30 frames per second, PAL operates with 25 frames per second and four fields. The frequency of the color subcarrier then turns out to be:

$$F_{sc} = \left(284 - \frac{1}{4}\right) F_L + 25 \text{ Hz} = 4.43361875 \text{ MHz}$$

The 25 Hz is an offset added for frequency compatibility with monochrome receivers. FL, of course, represents the line frequency, and ¼ is the difference between the field rate of the color system and the 50-Hz power line alternation.

There are at least three PAL demodulation schemes, from bare necessities to a deluxe version that even eliminates most recording errors. Simple PAL demodulation is about the same as R-Y and B-Y NTSC (perhaps two tubes, a pair of diodes, or two transistors) and chroma errors appear as changes in saturation. Standard PAL demodulators have a built-in delay line that offers many advantages including substantial elimination of both phase and saturation distortion. The third type, which probably appears only in high-priced television monitors, virtually eliminates all types of chroma and phase errors, especially time-base distortion. Here, each horizontal scanning line is regenerated with very close 4.43-MHz chroma sync over the entire line and all errors are carefully removed.

THE FRENCH SECAM SYSTEM

The SECAM abbreviation stands for Sequential With Memory, and was invented by Henri de France in 1957. It differs considerably from the NTSC system. Like PAL, however, it transmits R-Y and B-Y signals in sequence, but with a time lag of one line (64 microseconds) between them. Therefore, the French may use simple frequency modulation, instead of quadrature AM Modulation, to broadcast color information. R-Y is transmitted for one line then B-Y for the next, with color sync switching at the beginning of each field. In reception, the R-Y line has a positive sawtooth ramp that synchronizes the demodulator switching for satisfactory transmitter-receiver lock, and B-Y probably has the inverse. Therefore, a delay line is necessary in the receiver to retain the information of the first line while the set is reading the second line so the demodulator can operate on two lines together, decoding R-Y and B-Y simultaneously.

We are informed that there have been at least four versions of SECAM before it went on the air, with IVB being a version modified by the Russians, and introduced in Finland during 1968. Reportedly, it is closer to NTSC than previous SECAM versions.

SECAM I was the initial and simplest system with straight AM modulated R-Y and B-Y transmitted without sophisticated RF processing such as filtering, shaping, etc.

SECAM II synchronized an encoder and used FM modulation so that receiver decoders (demodulators) would lock at more precise frequencies and supply more accurate chroma information.

SECAM III was further modified to avoid image streaking and visible moire, plus certain incompatibility with monochrome transmissions.

SECAM IV, the 6.5 MHz system now in use throughout France and Russia, has been further developed with certain system time constants attenuated, along with modifications to the 4.43-MHz subcarrier, but it still transmits 1.5-MHz chroma sidebands. There is, however, R(red) pre-emphasis to increase energy in the lower and upper sidebands, which is then de-emphasized in the receiver to reduce color noise. It is expected that at a later date, the Russians will change over to SECAM IVB because of its added advantages over SECAM IV. It is also understood that in the IVB conversion, the Russians have further suppressed the subcarrier to remove its visibility on black-and-white reception and reinstituted the NTSC burst.

In regular European transmissions, our informant adds that the French are actually using the PAL system (at least in some instances) up to the transmitter, then coding PAL into SECAM for all French broadcasts, undoubtedly because of the large distribution of French receivers through the country.

OTHER WORLD STATISTICS

There are three other tables of interest that were made available to us by Television Digest's TV Factbook, 2025 Eye St., N.W., Washington, D.C. 20006. Table 16-3 is a listing of World television frequencies by channel numbers and video-audio carriers. And from the same publication, we've included the numbers of lines, channel bandwidths, and other vital statistics in Table 16-2, and the numbers and locations of the world's B&W and color television receivers in Table 16-4.

System	No. of lines	Ch. width MHz	Vision band-width MHz	Vision/sound separation MHz	Vestigial side-band MHz	Vision modulation	Sound modulation
A	405	5	3	−3.5	0.75	Pos.	AM
B	625	5	5	+5.5	0.75	Neg.	FM
C	625	7	5	+5.5	0.75	Pos.	AM
D	625	8	6	+6.5	0.75	Neg.	FM
E	819	14	10	±11.15	2	Pos.	AM
F	819	7	5	+5.5	0.75	Pos.	AM
G	625	8	5	+5.5	0.75	Neg.	FM
H	625	8	5	+5.5	1.25	Neg.	FM
I	625	8	5.5	+6	1.25	Neg.	FM
K	625	8	6	+6.5	0.75	Neg.	FM
L	625	8	6	+6.5	1.25	Pos.	AM
M	525	6	4.2	+4.5	0.75	Neg.	FM
N	625	6	4.2	+4.5	0.75	Neg.	FM

Field frequency: System M, 60 cycles per second, all other systems 50 cycles per second.

Picture frequency: System M, 30 cycles per second, all other systems 25 cycles per second.

Line frequency: System A, 10.125 kHz, systems E and F, 20.475 kHz, system M, 15.75 kHz, all other systems, 15.625 kHz.

Table 16-2. TV system characteristics. (Courtesy TV Digest)

Western European Standards

Channel	Vision Carrier (mc)	Sound Carrier (mc)[1]
E-2	48.25	53.75
E-2A	49.75	55.75
E-3	55.25	60.75
E-4	62.25	67.75
E-5	175.25	180.75
E-6	182.25	187.25
E-7	189.25	194.75
E-8	196.25	201.75
E-9	203.25	208.75
E-10	210.25	215.75
E-11	217.25	222.75
E-12	224.25	229.75
E-21	471.25	476.75
E-22	479.25	484.75
E-23	487.25	492.75
E-24	495.25	500.75
E-25	503.25	508.75
E-26	511.25	516.75
E-27	519.25	524.75
E-28	527.25	532.75
E-29	535.25	540.75
E-30	543.25	548.75
E-31	551.25	556.75
E-32	559.25	564.75
E-33	567.25	572.75
E-34	575.25	580.75
E-35	583.25	588.75
E-36	591.25	596.75
E-37	599.25	604.75
E-38	607.25	612.75
E-39	615.25	620.75
E-40	623.25	628.75
E-41	631.25	636.75
E-42	639.25	644.75
E-43	647.25	652.75
E-44	655.25	660.75
E-45	663.25	668.75
E-46	671.25	676.75
E-47	679.25	684.75
E-48	687.25	692.75
E-49	695.25	700.75
E-50	703.25	708.75
E-51	711.25	716.75
E-52	719.25	724.75
E-53	727.25	732.75
E-54	735.25	740.75
E-55	743.25	748.75
E-56	751.25	756.75
E-57	759.25	764.75
E-58	767.25	772.75
E-59	775.25	780.75
E-60	783.25	788.75
E-61	791.25	796.75
E-62	799.25	804.75
E-63	807.25	812.75
E-64	815.25	820.75
E-65	823.25	828.75
E-66	831.25	836.75
E-67	839.25	844.75
E-68	847.25	852.75

French Standards

Channel	Vision Carrier	Sound Carrier
F-2	52.40	41.25
F-4	65.55	54.40
F-5	164.00	175.15
F-6	173.40	162.25
F-7	177.15	188.30
F-8A	185.25	174.10
F-8	186.55	175.40
F-9	190.30	201.45
F-10	199.70	188.55
F-11	203.45	214.60
F-12	212.85	201.70

French Overseas Territories

Channel	Vision Carrier	Sound Carrier
F-4	175.25	181.75
F-5	183.25	189.75
F-6	191.25	197.75
F-7	199.25	205.75
F-8	207.25	213.75
F-9	215.25	221.75

Italian Channels

Channel	Vision Carrier (mc)	Sound Carrier (mc)
A	53.75	59.25
B	62.25	67.75
C	82.25	87.75
D	175.25	180.75
E	183.75	189.25
F	192.25	197.75
G	201.25	206.75
H	210.25	215.75
H-1	217.25	222.75

Eastern European Standards[2]

Channel	Vision Carrier	Sound Carrier
R-1	49.75	56.25
R-2	59.25	65.75
R-3	77.25	83.75
R-4	85.25	91.75
R-5	93.25	99.75
R-6	175.25	181.75
R-7	183.25	189.75
R-8	191.25	197.75
R-9	199.25	205.75
R-10	207.25	213.75
R-11	215.25	221.75
R-12	223.25	229.75

British Standards

Channel	Vision Carrier	Sound Carrier
B-1	45.00	41.50
B-2	51.75	48.25
B-3	56.75	53.25
B-4	61.75	58.25
B-5	66.75	63.25
B-6	179.75	176.25
B-7	184.75	181.25
B-8	189.75	186.25
B-9	194.75	191.25
B-10	199.75	196.25
B-11	204.75	201.25
B-12	209.75	206.25
B-13	214.75	211.25
B-14	219.75	216.25

Japanese Channels

Channel	Vision Carrier	Sound Carrier
J-1	91.25	95.75
J-2	97.25	101.75
J-3	103.25	107.75
J-4	171.25	175.75
J-5	177.25	181.75
J-6	183.25	187.75
J-7	189.25	193.75
J-8	193.25	197.75
J-9	199.25	203.75
J-10	205.25	209.75
J-11	211.25	215.75
J-12	217.25	221.75
J-45	663.25	667.75
J-46	669.25	673.75
J-47	675.25	679.75
J-48	681.25	685.75
J-49	687.25	691.75
J-50	693.25	697.75
J-51	699.25	703.75
J-52	705.25	709.75
J-53	711.25	715.75
J-54	717.25	721.75
J-55	723.25	727.75
J-56	729.25	733.75
J-57	735.25	739.75
J-58	741.25	745.75
J-59	747.25	751.75
J-60	753.25	757.75
J-61	759.25	763.75
J-62	765.25	769.75

[1] For channels E-21 through E-68, sound frequency is for systems G & H. For system I, image/sound separation, 6 mc; systems K & L, 6.5 mc.

[2] OIRT channels use Roman numerals, as OIRT-I, but this book uses R-1 instead because it is more consistent with other channel designations.

Table 16-3. World TV channel frequencies. (Courtesy TV Digest)

Country	Stations	Black & White	Color
Afars & Isis	1	1,500	
Albania	1	2,500	
Algeria	6	150,000	
Antigua (Low-power: 2)	1	6,500	
Argentina (Low-power: 33)	31	3,300,000	
Australia (Low-power: 37)	85	2,950,000	
Austria (Low-power: 68)	121	1,300,000	70,000
Barbados	1	17,500	
Belgium	19	2,100,000	5,000
Bermuda	2	18,000	
Bolivia	1	10,000	
Brazil (Low-power: 44)	50	6,500,000	
Bulgaria (Low-power: 84)	7	1,030,000	
Cambodia	2	18,000	
Canada	378	6,630,000	1,070,000
Chile	3	400,000	
China (Mainland)	30	300,000	
Colombia (Low-power: 11)	15	600,000	
Congo (Brazzaville)	1	1,500	
Congo (Kinshasha)	1	6,500	
Costa Rica (Low-power: 5)	4	120,000	200
Cuba	25	555,000	
Cyprus	2	42,000	
Czechoslovakia (Low-power: 300)	28	2,997,000	
Denmark (Low-power: 21)	30	1,475,000	35,000
Dominican Republic	6	77,000	
Ecuador	11	110,000	
El Salvador (Low-power: 1)	4	100,000	
Equatorial Guinea	1	500	
Ethiopia	1	6,500	
Finland	65	1,058,000	3,300
France (Low-power: 1,097)	107	11,000,000	400,000
French Guiana (Low-power: 2)	1	1,900	
French Polynesia (Low-power: 2)	1	3,700	
Gabon	2	2,400	
Germany (East)	25	5,300,000	10,000
Germany (West) (Lower-power: 1,158)	171	15,500,000	1,300,000
Ghana (Low-power: 1)	4	15,000	
Gibraltar (Low-power: 1)	1	6,100	
Greece	4	120,000	
Guadalupe (Low-power: 1)	1	7,200	
Guam	1	30,000	
Guatemala (Low-power: 3)	2	114,000	5,000
Guyana		3,500	
Haiti	1	10,500	
Honduras	5	45,000	
Hong Kong (Low-power: 8)	3[2]	429,000	
Hungary	11	1,596,000	
Iceland (Low-power: 25)	7	36,000	
India	1	21,000	
Indonesia (Low-power: 1)	4	75,000	
Iran	8	250,000	
Iraq	4	180,000	
Ireland	22	520,000	2,000
Israel (Low-power: 6)	12	340,000	
Italy (Low-power: 1,110)	72	9,700,000	
Ivory Coast (Low-power: 3)	4	11,000	
Jamaica	9	70,000	
Japan (Low-power: 962)	186	19,000,000	5,150,000
Jordan	1	50,000	
Kenya	3	20,000	
Korea (Low-power: 15)	7	400,000	
Kuwait	2	90,000	
Lebanon	9	250,000	
Liberia (Low-power: 1)	1	6,500	
Libya	2	200,000	
Luxembourg (Low-power: 3)	1	69,500	1,500
Malagaxy Republic (Madagascar)	1	1,000	
Malaysia (Low-power: 3)	12	180,000	
Malta	1	63,000	

1971-72 Edition

Country	Stations	Black & White	Color
Martinique (Low-power: 1)	1	9,500	
Mauritius (Low-power: 3)	1	20,000	
Mexico	62	2,500,000	175,000
Monaco	1	16,000	
Mongolia	1	600	
Morocco (Low-power: 5)	8	145,000	
Netherlands	13	3,000,000	150,000
Netherlands Antilles (Low-power: 1)	2	32,000	
New Caledonia (Low-power: 2)	1	5,000	
New Zealand (Low-power: 325)	20	625,000	
Nicaragua (Low-power: 1)	2	55,000	
Niger	1	100	
Nigeria (Low-power: 2)	8	53,000	
Norway (Low-power: 265)	64	850,000	3,000
Okinawa (Low-power: 1)	2	230,000	
Pakistan	5	80,000	
Panama	11	157,500	
Paraguay	1	25,000	
Peru	18	325,000	
Philippines	24	415,000	6,000
Poland (Low-power: 64)	24	4,023,000	
Portugal (Low-power: 23)	8	360,000	
Puerto Rico	—	600,000	
Qatar	1	20,000	
Reunion (Low-power: 4)	1	21,000	
Rhodesia	2	45,000	
Rumania (Low-power: 65)	17	1,289,000	
St. Pierre & Miquelon	1	1,500	
Samoa	1	3,000	
Saudi Arabia	8	300,000	
Senegal	1	1,500	
Sierra Leone	1	3,500	
Singapore	2	149,000	
Southern Yemen	6	22,000	
Spain (Low-power: 618)	31	4,050,000	1,000
Sudan	2	65,000	
Surinam (Low-power: 2)	1	30,000	
Sweden (Low-power: 71)	173	3,000,000	200,000
Switzerland (Low-power: 183)	83	1,226,000	62,500
Syria	5	118,000	
Taiwan (Low-power: 3)	4	420,000	
Tanzania	1	3,000	
Thailand (Low-power: 4)	5	225,000	
Trinidad (Low-power: 1)	2	54,000	
Trucial States	1	1,000	
Trust Territory of Pacific	1	1,500	
Tunisia	9	50,000	
Turkey	2	100,000	
Uganda	6	12,000	
United Arab Republic (Low-power: 10)	23	500,000	
United Kingdom	232	16,000,000	750,000
Upper Volta	1	3,000	
Uruguay	12	225,000	
USSR (Low-power: 698)	167	28,000,000	
Venezuela	29	700,000	
Vietnam (South)	1	25,000	
Virgin Islands	1	17,000	
Yugoslavia (Low-power: 182)	35	1,767,000	5,000
Zambia	3	18,500	
TOTAL	2,792	169,593,500	9,404,500
Satellites & repeaters	7,542		
U.S.	881	*61,400,000	*31,300,000
U.S. Military	49		
GRAND TOTAL	11,264	230,993,500	40,704,500

[1] Stations included in U.S. count. [2] One closed-circuit system.
* Preliminary estimate.

Table 16-4. World TV stations and receivers. (Courtesy TV Digest)

Answer Section

CHAPTER 1 ANSWERS

1. Electromagnetic waves
2. Same as 1.
3. Magnetic and electric fields.
4. Time and deviation (or amplitude)
5. In AM modulation, the carrier envelope is "expanded" or compressed by another signal. In FM, carrier amplitude is held constant while the carrier frequency is varied.
6. Upper and lower chroma sidebands
7. Same as 6.
8. 12 VHF and 70 UHF, a total of 82 channels, each with a passband of 6 MHz
9. 54 MHz to 216 MHz, and 470 MHz to 890 MHz
10. By 57 and 147 degrees, respectively
11. Same as 10; burst is 0 degrees
12. V, 1.4 milliseconds; H 11.1 microseconds
13. Same as 12.
14. The vertical and horizontal blanking intervals
15. 75 percent; 12.5 percent; between 25 and 30 percent, round figures
16. H, 59.94 Hz; V, 15,734.264 Hz; C, 3.579,545 MHz
17. 15,734 kHz, the horizontal scanning frequency
18. Y equals 30 percent red + 59 percent green + 11 percent blue
19. See Fig. 1-10.
20. Above by 4.5 MHz
21. To avoid excessive standing wave ratios (SWR) that can mean reflected signals which cause images and picture ghosts and distortions.
22. The 3.579,345-MHz subcarrier oscillator
23. They compress sync pulses and video levels, causing vertical or horizontal rolling and often an overly contrasty picture that may even go negative. Audio will often buzz with IF amplifier overload.
24. Basically, it's a frequency discriminator-to-DC converter operating at a center frequency of 45.75 MHz.
25. Negative; better AGC control, 30 percent extra output power, noise in black region always less noticeable.

CHAPTER 2 ANSWERS

1. Brightness—picture shadings, the gray scale; hue—describes the actual color; saturation—the degree of color intensity.
2. Because it adds luminance to the primary red, blue, and green colors
3. More than 30,000; between 400 and 700 nanometers at frequencies ranging from 3×10^{14} and 3×10^{15} Hz.
4. All three **unmodulated** saturated colors
5. Anything; same as 4
6. NTSC and gated rainbow
7. R-Y and B-Y signals, like I and Q, contain green, too, and are 90 degrees out of phase when demodulated, so green—210 degrees from R-Y

and 120 degrees from B-Y—is by no means a quadrature signal and is not useful.

8. Same as 7

9. The color subcarrier xtal; 0.0003 percent or 10 Hz in 3.6 million

10. Same as 9.

11. A field of non-contaminated color

12. 110 degree deflection angle; Chromacolor screen—in-line gun structure

13. A pattern of brightness variations appearing as sheer window curtains with overlapping folds. Actually, line-to-line variations as electron beams miss shadow mask holes. More prevalent because of better-designed tubes.

14. Dynamic convergence is already built in

CHAPTER 3 ANSWERS

1. EIA color-bar pattern, and the vertical interval test signal (VIT).

2. 195 divided by 80 or 2.44 MHz (something near the bandpass of many of the cheaper monochrome receivers).

3. High-frequency horizontal resolution

4. Ringing, regeneration, interference, receiver bandpass, etc.

5. The sweep-marker generator; f equals the number of lines divided by the sweep rate. However, you use this latter method only for very heavy input signals of millivolts to volts, which probably saturate the RF input.

6. To display the relative phase and amplitude of chroma signals on polar coordinates.

7. Look at both the EIA color bar broadcast pattern and the VIT signal.

8. Yes, when both are included in receiver controls; noise is adjusted first.

9. It fine tunes the local TV oscillator and can move a receiver's video, sound, and chroma response around on the response curve, causing much interaction.

10. No, Zenith's new solid-state set doesn't.

11. Vertical height and linearity, sometimes HV centering

12. Set for minimum of color snow in the raster on any blank channel.

CHAPTER 4 ANSWERS

1. One or more beams are not striking the color phosphors as they should.

2. Gun beams striking other than normal dots and in a weak magnetic field.

3. Back toward the convergence assembly.

4. Before

5. Before

6. To agitate and randomly orient magnetic lines of force in the metal shadow mask (demagnetize).

7. False; it increases

8. 100 percent

9. The screen cutoff points so all guns will extinguish at the lowest brightness levels.

10. Parabolic

11. Sawtooth

12. Two-pole convergence with red-green lateral magnets, simplifying red-green convergence.

CHAPTER 5 ANSWERS

1. Each marker is at 50 percent points on curve

2. The 41.25-MHz sound trap and 47.25-MHz lower-adjacent sound trap.

3. Same as 2.

4. Sound by 4.5 MHz

5. False. Almost all UHF tuners use semiconductors.

6. Positively not! The graph in Fig. 5-4A is substantial proof, at least from this manufacturer.

7. Low noise, adequate bandpass, good adjacent-channel sound rejection, low SWR, etc.

8. Selecting certain RF frequencies, then converting separate video and sound carrier to IF frequencies.

9. The internal generation (modulation of one by the other) of extra RF signals from two strong incoming signals.

10. Take your answer from Fig. 5-4. Of course, it is yes, since there's no filament, less heat, low noise, good gain, etc.

11. Basically, the turret tuner has channel strips mounted on a rotating drum instead of wafers, plus individually tuned RF and oscillator sections instead of series coils shorted by switches.

12. By application of DC voltages that make the diodes act like capacitors.

13. Amplifies video and chroma IF carriers for chroma, sync, and luminance processing by the rest of the receiver circuits.

14. Because of possible interference between it and the chroma information, which would cause chroma desaturation in the picture.

15. See Table 5-2.

16. Rolloff.

17. Video IF, sound, and chroma (represented by sidebands)

18. No, because the LA trap is on the opposite side of the response curve from the 41.25-MHz sound carrier, at 47.25 MHz.

19. They consumed power, drew large currents, and would break down easier then less loaded resistors.

20. To achieve broader bandpass response and trap suckout.

21. RCA

22. This question is open to discussion. Consider the progress and multiple functions on a chip, then realize what the old systems did NOT do. Base your arguments on utility and not parts count; also, remember that the IC is mounted on a quickly detached module.

23. 1) No; 2) markedly. The RCA IC looks much different from the Motorola discrete component system.

CHAPTER 6 ANSWERS

1. White to the point of blooming.

2. In the absence of signals, the CRT is dark and there is no excess current drain.

3. No, and therein lies a problem.

4. So that chroma and luminance (fine detail) can be processed together when they are matrixed either before or in the picture tube. Delay is 0.8 microseconds.

5. In circuit with a dual trace, triggered sweep, carefully calibrated oscilloscope.

6. In the luminance amplifiers.

7. It is cutoff

8. Nonlinearities that look like spurious voltages developed by nonlinear, half-wave diode video detectors.

9. By using a double-balanced, full-wave video detector; filtering is another.

10. 50 kHz to 150 kHz, a ratio of 3:1.

11. Foster-Seely discriminator

12. Yes, they're usually better designed and have excellent bandwidths.

13. Parallel, series

14. The outputs. The ratio detector output is a ratio change across the output resistors, while that of the discriminator is a voltage difference.

15. Its tuned quadrature grid causes larger or smaller current phase lags as a result of incoming signals which produce different pulse current widths at the detector plate.

16. Complementary symmetry output stage and capacitor coupling to the speaker.

17. Positive forward bias on the emitter, since its base is already biased on by R6 and R8.

CHAPTER 7 ANSWERS

1. To time the horizontal and vertical oscillators.

2. A diode. Today it's a separate circuit driven by the video amplifier and keyed into conduction by a pulse from the flyback transformer.

3. Low voltage, selected load and bias control.

4. Positively, and this early module control (it has since been eliminated) would cause the receiver to lose sync if it is not set correctly.

5. In microseconds, so the disturbance will affect only one part of a line.

6. Same as 7.

7. They show how exceptional IF amplifiers, if designed properly, can peak on weak signals in both video circuits and at the sound detector.

CHAPTER 8 ANSWERS

1. $1/59.94$ Hz; two; 16.664 msec.

2. Interaction of linearity and size (height) controls, plus all around good linearity.

3. Trapezoidal waveform; a sawtooth

4. Resistors and capacitors

5. Because of its input DC-set RC time-constant and feedback from the vertical output stage.

6. Better frequency stability with temperature, small size, lower cost, and greater linearity with fewer components.

7. In Motorola it's 5.33 ohms at f_0.

8. With a signal in, no signal out and DC voltages OK, substitute a complete IC for a cure if the grounds (chassis common) are secure.

9. C708, C709

CHAPTER 9 ANSWERS

1. They must have good hold and pull-in ranges, proper phase differences, little noise, good damping, and proper time constants.

2. At turn-off time when high voltages and currents coincide, causing the highest collector dissipation.

3. Horizontal repetition rate; because the raster is a 525-line system and all the lines must be generated in the time of one 30th second.

4. Yes. RCA uses a pair of SCRs, many use brightness limiters, there are HV doublers, triplers, quadruplers, and all these have smaller and in many cases unshielded flybacks.

5. They are RC series time constants in the control section of any horizontal oscillator to prevent over control when the receiver is warming up.

6. It is the anti-hunt network.

7. At ¾ths of the rising forward scan.

8. Being a voltage-sensitive resistor, its resistance decreases with applied voltage and so reduces the grid voltage and the output tube's ability to conduct.

9. 1) Trace resonant, 2) Retrace resonant, 3) Power resonant

10. In RCA semiconductor sets, in parallel.

11. By cutting off its current, or reducing the anode-cathode voltage.

12. Same as 11

13. The retrace SCR conducts for some 25 microseconds, while the trace is on for only about 15 microseconds. This is because the trace moves the scan beam only about ³₈ths of the time, or 26 microseconds, vs 63.5 microseconds for retrace and blanking.

CHAPTER 10 ANSWERS

1. Have them oscillator driven and well-regulated.

2. To provide better current regulation and give the capacitive aquadag coating on the picture tube a chance to discharge. If it didn't discharge, a brightness spot would remain on the CRT long after the receiver was turned off.

3. To supply the deflection yoke with a linear sawtooth current to move the trace across the screen.

4. Absolutely not. The grid comes on first as shown in Fig. 10-3.

5. In the same direction.

6. It increases proportionally, if the control and circuit are linear.

7. Initially, it decreases, but then tries to increase. Certainly, if there's too much beam current without brightness limiting, there will be too much HV current drain and the flyback could go up in smoke.

8. How about the luminance amplifiers? A lot of service people always miss this one.

9. Oscillator drive of all voltage supplies, especially low voltage.

10. Center and amplitude.

11. The yoke current is shaped to linearize the sweep, not selective beam bending as with dynamic convergence.

12. Let R333 equal Rx, then:

$$10 = 200 \times \frac{Rx \times 10}{10^6 + Rx} , \quad Rx = \frac{10 \times 10^6}{1.990 \times 10^3}$$

$$5.03 \times 10^3 \simeq 5K = \text{answer.}$$

13.

$$12 = \frac{10^3}{16^6 + 10^3} = \frac{12 \times 10^6 + 12 \times 10^3}{10^3} = x \simeq 12 \times 10^3$$

$$\text{Ein} = 12 \text{ KV}$$

14. Control beam current as a function of high voltage.
15. Magnetic flux
16. Replaced by the retrace SCR.
17. Due to the wider deflection angle of the 110-degree tube.

CHAPTER 11 ANSWERS

1. A set with a power transformer has isolation from house current; because it's less expensive than using a power transformer.
2. Through D1, which conducts on a negative input; D2 conducts on a positive input.
3. RMS x 1.414 equals peak, and 2 x 1.414 equals peak-to-peak. So 340 divided by 2.828 equals 120 volts, RMS, the usual receiver design center.
4. Less heat, reliable, maintain life efficiency, require smaller power transformers.
5. It becomes an AC shunt.
6. Voltages drop and AC ripple appears.
7. You don't! Only a totally open or shorted reactance can be accurately evaluated. You must use current, for instance, to evaluate electrolytic capacitors, or a capacitance checker, and a pulse voltage and an oscilloscope for inductors.
8. Base and emitter (currents)
9. gm equals u/rp for the tube, while gm equals Id Vgs for the FET; AV equals gm x ZL which translates to gm equals AV/ ZL.
10. To comply with FCC regulations to disable part of the receiver in case the high voltage became excessive and threatens X-radiator.
11. By modulating the B+ supplied to the horizontal output transformer with a 60-Hz parabola.
12. By using regulators in the individual modules and not in the main supply.
13. AC all the way. If you missed this one, re-read the theory of operation twice!
14. The oscillator in Delco's is a 20-kHz free-running Colpitts instead of a "governed" oscillator at 15,734 Hz.

CHAPTER 12 ANSWERS

1. Color, burst, and flyback gating pulses
2. At or after the demodulators. In old-style receivers it is not used in the color subsystem. Instead, luminance is matrixed with chroma in the cathode ray tube.

3. Keeps the oscillator in color sync and somewhat controls the gain

4. Always externally

5. Down below the X axis in the fourth quadrant

6. 85, 90, and 105 degrees

7. In the cathode ray tube between control grids and cathodes.

8. From the negative outputs of the R-Y and B-Y demodulators across a resistive matrix.

9. No qualifications are required in the demodulation process to compensate for gray-scale tracking adjustments made solely in the luminance channel.

10. The 3.579,545-MHz chroma subcarrier.

11. 33 degrees

12. 3.08 MHz to 4.08 MHz; I and Q, since the R-Y and B-Y phase shift has not yet theoretically occurred.

13. The color killer

14. Traces of sync, slanting white lines, or flashes of color could appear on the CRT screen. One receiver with shorted vertical guide diodes even produces horizontal traces during this interval.

15. By the amplitudes of the demodulated RGB outputs

16. Injection lock; the sync burst rings the crystal directly.

17. It's a Schmitt trigger; one half is off when the other half is on.

18. Three, since IC1 is a tri-phase demodulator.

19. Delivers more red to the CRT and changes the demodulator phase angle to include broader fleshtones.

20. Accu-Tint (RCA)

21. Preset (passive) brightness, hue, contrast, and intensity, while active semiconductors increase the red output gain, ride DC gain on the second color IF, and change the demodulator phase.

22. Switching, since the chroma is demodulated by 3.58-MHz switching at different phase angles.

23. No transformers to adjust, only 3 DC potentiometers

24. A color-bar generator and a vectorscope

25. Certainly. With no output from no. 1, there's no input to no. 2; therefore, no combined output.

26. No. A great deal of chemical and mechanical progress has been made since the original specification.

27. The latter. It's more expensive sometimes to generate, but it's also very reliable.

28. Be careful of neutralization and damping so the circuit has maximum transient response.

29. It will lock on a single cycle of burst.

30. "...out of phase with those opposite, so one pair switches while the other pair is off."

31. Differential amplifiers

32. A Schmitt trigger color killer

CHAPTER 13 ANSWERS

1. 1.5 percent

2. Shielded cable, preferably 300 ohms

3. No! Overall losses in all cables are greater at higher frequencies.

4. Make at least a visual inspection of the antenna and lead in.

5. Because of its characteristic impedance match with the receiver and the antenna. It should also contain the surrounding magnetic field in

its encapsulated sheath, and have the correct wire size and conductor spacing.

6. When you have seen the gain and lobe patterns and know the guaranteed strength in high winds

7. Normally, such an antenna doesn't exist.

8. See Fig. 13-1. They must be separated enough to avoid interference with each other.

9. They may turn or jump out of the connecting joints in high winds if you don't.

10. They last, look pretty, and you get more money for your installation.

11. No, it will absorb water and dirt, deterioriate, and lose signal and add ghosts. Furthermore, don't use ordinary twinlead except for FM radios.

12. Leading ghosts are normally found in installations close to the transmitter, while trailing edge ghosts are encountered in more remote locations—some distance, that is, away from the transmitter.

13. Use a fully shielded line and a directional, narrow front lobe antenna and a rotor, if at all possible.

14. 60 to 80 db.

CHAPTER 14 ANSWERS

1. Have a need where the instrument will pay for itself.

2. Perhaps to measure high voltage, current, and ohms. But there are even scopes by Tektronix that will do all these things and many more. The realistic answer is no.

3. Digital multimeters, many of which already have astonishing accuracies, good ranges, and sell for less than $400.

4. Simply flip the AC-DC switch to DC and count the volts (determined by the number of graticule divisions) the waveform rises or falls.

5. Certainly, because at DC, the scope reads waveforms at peak-to-peak values anyway, just as it does on pulse voltages. And pulse voltages of certain repetition rates and durations are equal to DC voltages. Sine waves, however, must be divided by 2.828 to equal an RMS-DC calibration.

6. 2 to 5 milliseconds and 10 to 20 microseconds per division.

7. Yes, since it's taken at the AFC diodes. Sync information here is constantly correcting the horizontal oscillator.

8. The chop rate so far in oscilloscopes just isn't fast enough. Rarely does it exceed 150 kHz to 200 kHz, if that. So you'd see chopping "hash" all over the place and there would not be a useful signal display.

9. You might make a mistake among the changes, and this could put you in double trouble, not to mention the time wasted.

10. Not like the vacuum tube receivers where much of this trouble occurred in the horizontal output tube between grids and plate in the form of small but persistent oscillations. Semiconductor sets certainly develop faults—and some may look like Barkhausen, but they're really not.

11. 1) Saturable reactors; 2) solid state

12. Birdie, absorption, pulse and intensity.

13. Probably not, but a trap might be misadjusted or the fine tuning may not be right.

14. Probably not. The problem is associated with one of the video amplifiers.

15. You probably **do** need alignment. A test pattern or VIT signal would undoubtedly tell you right away.

16. Because you're using an oscilloscope as a real-time X-Y plotter and neither the receiver nor oscilloscope need vertical and horizontal sync. The sweep generator, however, does produce a 60-Hz sine wave for the X-amplifier terminals of the oscilloscope.

17. The double markers are the large ones and the f_0 resonant frequency output is exactly between them.

18. The vectorscope can see only R-Y, B-Y; not G-Y and luminance.

19. Not always, especially if the positioning is critical.

20. Insufficient AGC bias, a bad cable connection or termination, or a bad sweep generator.

21. Find a common, secure chassis point, and make sure all equipment grounds meet at this point.

22. Turn off the line markers one by one and you'll know which ones to count on.

23. Not usually, and their frequency characteristics are lousy.

24. A triggered sweep oscilloscope with an accurately calibrated time base.

25. Because there may be trouble in a luminance amplifier instead, and you need to know if the tube emission is OK, whether it has shorts, and if its three guns track. A "gassy" tube might just turn up good. Of course, a gas check provision in the tube checker might help, too.

Index

D

E

F

G